THE COMPLETE BOOK OF CAR MAINTENANCE AND REPAIR

THE COMPLETE BOOK OF CAR MAINTENANCE AND REPAIR

a survival manual

JOHN D. HIRSCH

Illustrations by Lionel Sacks and Frances Johansson

CHARLES SCRIBNER'S SONS · New York

Library of Congress Cataloging in Publication Data

Hirsch, John D
  The complete book of car maintenance and repair.

  1. Automobiles—Maintenance and repair. I. Title.
TL152.H54   629.28′8′22   73-1355
ISBN 0-684-13400-4

1 3 5 7 9 11 13 15 17 19 V/C 20 18 16 14 12 10 8 6 4 2

Printed in the United States of America

For rearing me, correcting my grammar, and for help with this book in many ways, I owe thanks to my mother, Amy B. Hirsch.

Encouragement and valuable suggestions were offered by Ed Feldmann, Max Gartenberg, Barry Bruce-Briggs, and Daniel Tanner.

Illustrations have been credited individually and are marked as follows:

"Ford"—courtesy of Ford Customer Service Division.

"Dodge"—courtesy of Chrysler Motors Corporation.

"Chevrolet" and "Pontiac"—courtesy of those respective Divisions of General Motors.

"Volkswagen"—courtesy of Volkswagen of America.

"Snap-On"—courtesy of Snap-On Tools Corporation.

"Owatonna"—courtesy of Owatonna Tool Company.

The tables on the cost of operating an automobile on pages 333 and 336—courtesy Office of Highway Planning, Highway Statistics Division, U.S. Department of Transportation.

Information was supplied by the Hand Tool Division of Dresser Industries (makers of S-K Tools), Mac Tools, Oldsmobile Division of General Motors Corporation, and Compu Industries, Inc.

Any errors there may be in this book can be blamed on the little gremlins that bedevil authors. Despite all efforts to eradicate them, they do manage to pop up now and again.

Contents

Introduction

Some people never seem to have any luck with their cars. A wheel comes off, the engine starts to overheat and requires major repairs, or maybe the brakes give out and cause an accident. Obviously the car is a lemon.

While there's no denying that some cars *are* lemons, the owner who has too many bad cars is suspect. No car can stand up under poor driving techniques, lack of maintenance, and sheer ignorance of basic car components and how they function.

Take the case of a 1969 Chevrolet which had an exhaust pipe bracket rusted through. Incredibly, the owner ignored the rattle and clatter of the loose exhaust until the whole muffler and exhaust assembly fell off. A little $1 repair escalated to one costing $40 or more.

But that's not the end of it. If the muffler and exhaust assembly is not replaced it can lead to more trouble—anything from engine failure to carbon monoxide poisoning.

This is not to say that a faulty exhaust system will lead to burned-out valves, or that poison gases will find their way into the driver's compartment, just that it *could* happen and has. Better to replace the $1 exhaust bracket, even if it causes a little inconvenience.

The key to economical and trouble-free operation of a car is regular maintenance plus enough knowledge to spot developing problems.

If you are very lucky you may be able to find a trustworthy and skillful mechanic who will look after your car as if it were his own. Such mechanics are increasingly hard to find and are usually overworked. Their labor charges will run from $7 to $10 an hour or more and parts are priced at their full retail value.

You will save money, time, and frequently a lot of aggravation by learning to do minor maintenance and repairs yourself. You do not need any special experience or even a mechanical bent to get started. All it requires is a little effort and the willingness to devote a few spare weekends to working on your car. Any man or woman can do it.

Aside from the money you save, the big advantage of doing repairs yourself is learning enough so that you can deal with emergencies. When your car won't start on a cold morning you will be able to take advantage of simple troubleshooting procedures rather than having to wait until the local garage gets around to you.

Even learning the most basic operations can save you a lot of money. For instance, by buying oil in bulk and adding it yourself, you can save 30 or 40 cents a quart over the gas station price. This adds up to a substantial amount during the life of a car.

Then, if you can change your own oil and oil filter—a job which can be done in a spare half hour with only a few tools—there is a further saving of as much as five dollars each time the job is done.

Changing an air filter is even simpler. To replace a dry element filter (the kind on most modern cars) just loosen the top bolt(s), pull off the hoses (on a car with emission controls), and open up the filter canister. Removing the old filter and installing a new one is a job which takes about two minutes. The garage price for changing an air filter is around $4. By purchasing a filter in the auto department of a discount store, you can save at least half this price. Two dollars for two minutes, not bad?

Chapter 1 of this book is all about how your car works. The next five chapters cover maintenance and repairs.

Chapters 7, 8, and 9 deal with roadside emergencies and troubleshooting —everything from getting your car started when it won't run to getting it unstuck from axle-deep mud or snow.

In Chapter 10 major repairs are considered from two different viewpoints—doing it yourself or dealing with a professional repairman and not getting scalped. This chapter is intended to supplement, not replace, a service manual giving a step-by-step guide to major repair procedures on one particular car. It will help you understand and interpret the more technical shop manuals.

The final chapter is all about purchasing, financing, and insuring a car. There's a guide to checking out the condition of a used car, and advice on how to sell your old one for the best price.

In the appendices you'll find some handy lists of tools to buy and places to get parts and service manuals for almost any car. There's also a comprehensive review of maintenance procedures and when they need doing.

This is not a book for hot rodders, master mechanics, or millionaires. It's designed for people who drive old Ford Falcons and anxiously listen to the engine clattering, hoping the car will last at least another year.

Above all, it's a book to be used as well as read. You will not be getting your money's worth unless the pages wind up with at least a few greasy fingermarks.

So step outside and bring along a few tools. The time to begin getting acquainted with your car is now . . .

1

Giving Your Car a Once-Over

● Begin on a quiet Sunday when the weather is warm. Raid the family tool box for an assortment of wrenches and screwdrivers. You won't need any special tools, but you should have a tire pressure gauge and some blocks or stands to put under the car when you want to work beneath it.

Your first good look under the hood of the car may be confusing. Surrounding the engine there are pumps, electric motors, wires, hoses, drive belts, filters, and other strange devices. You will understand these components better if you look at those that work together as a system to accomplish specific goals.

The systems of a modern car are:
- *the engine*, which produces power by burning fuel;
- *the fuel supply system*, which delivers gasoline mixed with air to the engine cylinders;
- *the exhaust system*, which carries away the waste gases left after fuel is burned in the cylinders;
- *the ignition system*, which creates the spark needed to ignite the fuel mixture in the cylinders;
- *the starting system*, which turns the engine crankshaft to set the whole cycle in motion;
- *the charging system*, which generates the electric power needed to keep the battery charged and run the lights, horn, and other accessories;
- *the cooling system*, which dissipates engine heat;
- *the engine lubrication system*, which circulates oil to the moving parts of the engine;
- *the clutch and transmission system*, which transmits engine power to the drive train and gives the engine a mechanical ad-

vantage by varying the number of times the engine crankshaft must turn to rotate the drive wheels once;

- *the drive train,* which brings the power to the drive wheels (on most cars the drive wheels are the rear wheels);
- *the braking system,* which slows or stops the car;
- *the steering system,* which turns the front wheels of the car;
- *the suspension system,* which connects the car frame to the wheels (it has springs and shock absorbers to provide a comfortable ride and good handling).

All these systems are part of a typical car, but the way they work may vary. If you have a Volkswagen, for instance, there is no radiator and water pump circulating coolant to the engine. Instead the cooling system has a fan and shrouding to circulate air around the engine.

On some cars the engine powers the front wheels instead of the rear wheels. Others have an automatic transmission in place of a clutch and gearbox. A few cars have their engines in the rear and a gearbox in unit with the rear axle.

Let's examine your car together to see how it works. We will combine some theory with the identification of component parts and a few basic maintenance procedures to keep the car running well.

Put on some old clothes, a cloth cap to cover your hair, and use waterless hand cleaner to take the grease right off when you are done.

THE ENGINE

Open the hood of your car and look at the engine. You cannot see the internal parts of the engine, but the block and head are readily visible. The cast-iron or cast-aluminum cylinder block is the main part of the engine and the head is bolted to it, covering up the cylinders. A six-cylinder engine has six cylinders or holes cast into the block, while a V-8 engine has two banks of four cylinders each.

In each cylinder there is a snug-fitting piston which can ride up and down, just touching the cylinder walls. As the piston moves upward in the cylinder, it compresses a mixture of gas and air. At the crucial moment when the gas/air mixture has been compressed about as much as possible, a spark is introduced into the cylinder and the mixture explodes. The force

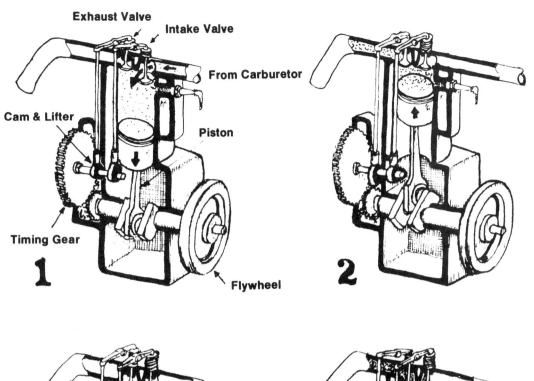

Exhaust Valve

Intake Valve

From Carburetor

Cam & Lifter

Piston

Timing Gear

Flywheel

1

2

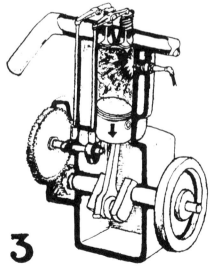

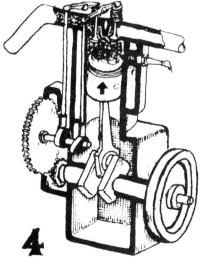

3

4

*Operation of four-cycle piston engine: (1)in-
take (2)compression (3)combustion (4)exhaust.*

3 · Giving Your Car a Once-Over

of the explosion causes the piston to be driven downward. Each piston is connected to a heavy rod called the crankshaft. As the pistons are driven downward in turn, the pushing force behind them goes into rotating the crankshaft.

At the side or top of each cylinder there are two valves which can open and shut. One is the intake valve which opens to let the fuel mixture in. The other is the exhaust valve which releases the burned gases after combustion. There is also a spark plug screwed into each cylinder to supply the spark which ignites the mixture.

The common gasoline engine is known as a four-cycle engine. This is because only one out of every four piston strokes produces power.

A complete cycle begins with the piston at the top of the cylinder. It moves downward creating suction and, at the same time, the intake valve opens. The fuel mixture is sucked into the cylinder. This is known as the *intake stroke.*

Next is the *compression stroke.* Both intake and exhaust valves are closed as the piston moves back upward, compressing the mixture. Compression is necessary because the mixture will explode with greater force when confined.

Then comes the *power stroke.* Both valves are still closed as the compressed mixture is ignited by a spark from the spark plug and explodes, sending the piston downward again with great force.

Finally there is the *exhaust stroke* when the piston moves upward again and the exhaust valve opens so burned gases can be forced out.

The whole cycle is repeated in each engine cylinder many times a minute. When it is stated that an engine is turning over at 5,000 revolutions per minute (r.p.m.), this does not mean that the whole engine is physically turning over but simply that the pistons in each cylinder, igniting in order, are turning the crankshaft this many times per minute.

The crankshaft, to which each piston is attached, develops the engine's power and translates it to a rotary motion. Each piston lends movement to the crankshaft during the piston's power stroke. The crankshaft moves the piston during the other three strokes.

The crankshaft is linked to another rod called the *camshaft* by a timing chain or gear. The crankshaft turns the camshaft and the cam causes the valves to open and shut at the right time.

Even the spark is controlled indirectly by engine rotation. A gear on the camshaft runs the distributor which sends voltage to each of the plugs in sequence.

The parts of the engine are linked as closely as the hipbone and the thighbone. The motion of one component lends motion to another. And the whole object is to keep the crankshaft rotating so the force of its movement can be delivered to the wheels of the car.

Naturally the engine must be kept properly lubricated. Oil is generally kept in the crankcase, although some cars have a separate reservoir for oil. A pump forces the oil to the moving parts of the engine.

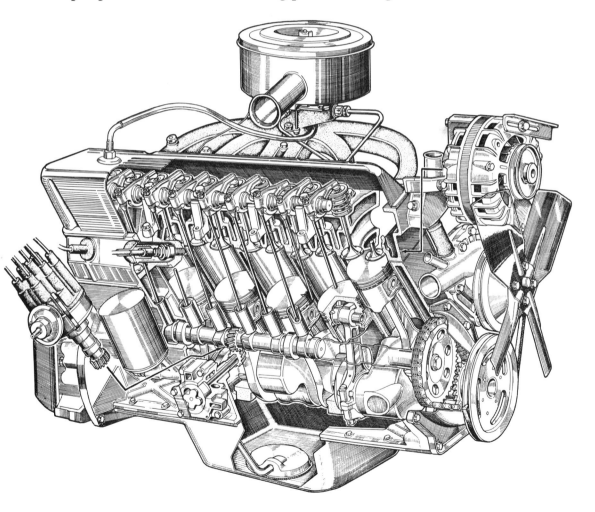

Cutaway view of 6-cylinder engine. (Dodge)

Check the oil level by locating the dipstick on one side of the engine. The oil should be checked while the engine is cold.

Most dipsticks have a top mark indicating that the crankcase is full and another mark below it which shows when exactly one quart of oil is needed. Remove the dipstick, wipe it clean with a rag, then push it down as far as it will go. Some dipsticks require a little twisting to reach bottom. Any difficulty reading the oil level can be avoided by holding the dipstick against a sheet of white paper or a clean rag.

Checking oil level. (Volkswagen)

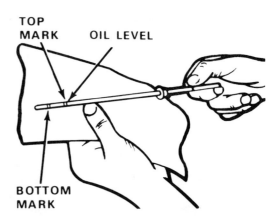

IGNITION SYSTEM

Find the spark plugs screwed into the engine block. The heavy wires from each spark plug lead back to an octopus-like device called the *distributor*.

The purpose of the distributor is to send a high-voltage impulse to each spark plug in turn, just as the mixture in the cylinder is compressed and ready to be ignited. This high-voltage current jumps from one electrode of the spark plug to another and creates the spark in bridging this gap. The same sort of spark is created when you bring a high-voltage wire near an electrical conductor and a spark leaps between the two.

Since the battery produces low voltage (6 or 12 volts) and the spark plug requires as much as 20,000 volts for a spark to leap the gap, voltage must be stepped-up somewhere. This is done in the *coil*.

The coil and distributor are partners and function together. Low voltage

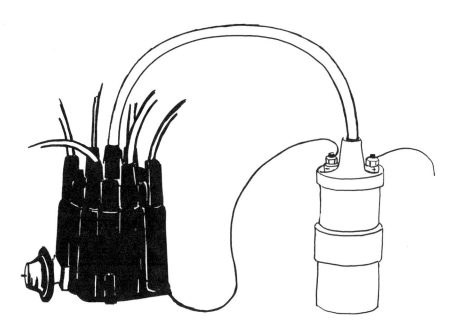

Coil and distributor. (Volkswagen)

goes to the coil whenever electrical contacts in the distributor (called *points*) are closed to complete an electric circuit. High voltage is created in the coil and returns to the distributor via a thick wire running from the top of the coil to the center of the distributor cap. Trace this thick wire from the distributor cap and you will find the coil, which looks like a metal can with a nipple on it.

When high voltage goes back to the top of the distributor, it is fed to a rotating arm called the *distributor rotor*. As the rotor revolves it contacts metal terminals in the top of the distributor cap and sends the high voltage to each spark plug in turn, creating the spark which ignites the mixture in the cylinders.

In a sense the distributor is the "nerve center" of the car. It controls the sequence of combustion in the cylinders so that engine operations run smoothly. When the engine begins to run rough, it is usually parts of the distributor which must be replaced or adjusted. The ignition system will be described in more detail in Chapter 6.

To examine the parts of the distributor, remove the distributor cap by flipping off the clips at either side with a screwdriver. (Some caps have retaining screws rather than clips.) Leave the spark plug wires attached.

Do you see the rotor on the distributor shaft? Look for the condenser, which resembles a small battery, held to the distributor body by a retaining clamp, and the electrical contacts (points), which open and close rapidly as the engine runs. Have someone crank over the engine while you watch the rotor turn and the points open and close. Naturally the car will not start since no voltage is getting to the spark plugs. Replace the distributor cap the same way it came off.

STARTING SYSTEM

The battery supplies the current to start the car and get the whole works perking. Most batteries are located in the engine compartment, although a few are cleverly hidden away in the trunk compartment or under one of the seats in the passenger area.

Batteries have separate cells producing two volts apiece. A 12-volt battery has six cells and each has its own opening to add water.

Batteries are filled with water which mixes with acid to become the battery electrolyte. To avoid harmful impurities (such as iron) in some water, distilled water should always be used to fill the battery cells. Buy it from a supermarket or drugstore, or use a device sold to distill water for steam irons.

CAP

RETAINING CLIP

MARK

Opening distributor. (Volkswagen)

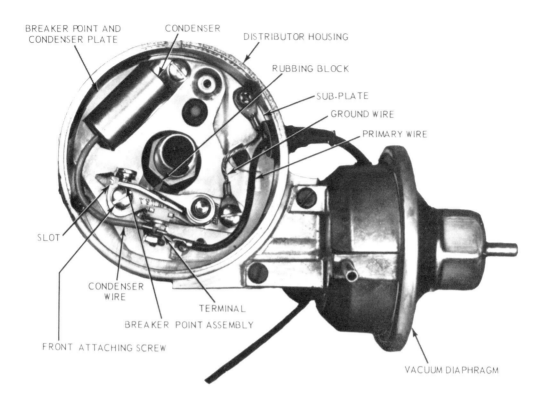

BREAKER POINT AND CONDENSER PLATE

CONDENSER

DISTRIBUTOR HOUSING

RUBBING BLOCK

SUB-PLATE

GROUND WIRE

PRIMARY WIRE

SLOT

CONDENSER WIRE

TERMINAL

BREAKER POINT ASSEMBLY

FRONT ATTACHING SCREW

VACUUM DIAPHRAGM

Points, condenser, and vacuum advance. (Ford)

Check each battery cell for water level. The water should be about one-quarter inch above the battery plates visible at the bottom of each cell. Most batteries have a round or star-shaped opening on the filler tube to indicate proper level.

Do not pour water into the cells from a can or bottle since it is too easy to overfill them. Use a rubber-bulb syringe or plastic squeeze bottle which is refillable.

On top of the battery are two terminals or posts with a cable clamped to each (a few new-model batteries have side terminals). There is a positive and a negative terminal, usually marked with plus and minus signs.

One cable supplies the current. It goes to the starter switch and branches off to supply current to other devices which run on electricity in the car.

General Motors cars have two current-supplying cables from the battery terminal. One goes to the starter switch and the other goes directly to a hook-up with accessory wires.

The cable from the other battery terminal is a *ground cable*. A ground is simply an electrical conductor which has no charge in itself. A wire or cable running to a ground is necessary to complete all electric circuits. The battery cable connects to a frame member or the engine block for a ground. Some ground cables are split and connect to both.

Most cars have the cable from the negative terminal grounded. Some

New battery design has one cell cap which glows when water is needed. (Chevrolet)

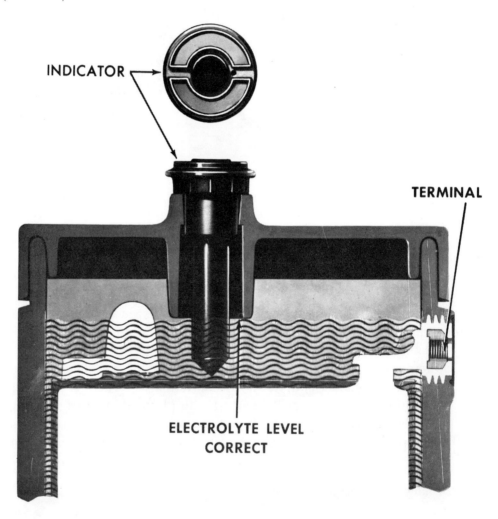

INDICATOR →

TERMINAL

ELECTROLYTE LEVEL
CORRECT

ONE PIECE COVER LEVEL INDICATOR VENT PLUGS

SEALED TERMINAL

Some batteries feature sealed side terminals. (Chevrolet)

English cars, early Volkswagens and Fords prior to 1956, are positive ground with the negative terminal supplying the power.

Trace the power-supplying cable from the battery to the starter switch or *solenoid*. The solenoid is a special kind of magnetic switch which generates more force than the conventional magnetic switch or *relay*. A moveable iron plunger is pulled into the solenoid's core by electromagnetism and pulls the starter drive into engagement. GM and Chrysler cars both have a solenoid mounted directly on the starter. In addition, Chrysler products use a simple relay mounted on the firewall (the wall between the engine and the passenger compartment) to pass current along to the solenoid. Ford cars have a different starter design which requires only a relay. It is mounted on the sidewall or firewall of the car.

The starter itself is commonly mounted low and toward the rear of the engine. If your car is a V-8 you may have difficulty seeing or reaching the starter.

The starter turns the engine flywheel, a heavy wheel attached to the back of the crankshaft. A pinion gear on the starter pulls into engagement with a ring gear on the flywheel and rotates, turning the crankshaft. Then the pistons and valves move and the engine cycle begins.

When the engine starts, the starter pinion gear retracts. The terrible noise you heard if you ever attempted to turn the ignition key to the start position while the engine was running is the pinion trying to engage a flywheel gear which is already rotating. Don't do it again as you could easily strip the flywheel gear.

GENERATING SYSTEM

The battery creates electricity through chemical action. Sulphuric acid in the battery reacts with lead plates in each cell to set up a flow of electrons within the plates. However, the acid soon gets absorbed in the plates and the battery goes dead.

For a battery to keep on producing power, it must be recharged frequently. The charge is nothing more than electric current which goes to the cells and drives the acid back out of the lead plates.

The current needed to keep the battery charged is provided by the *generator*. This is a device for turning mechanical energy into electrical energy. The generator is run by a belt driven by a pulley which is attached to the front end of the crankshaft. This belt is known as the *fan belt* since it also runs the radiator fan on most cars.

To spot the generator look behind the radiator for the fan and fan belt, then look for the generator at the other end of the belt. On newer cars the generator has been replaced with a similar but more efficient device known as an *alternator*. A generator is tubular and solid while an alternator is like a swollen pancake in shape and has cooling vents on the exterior shields. The differences in operation between a generator and alternator will be discussed in Chapter 4.

When the engine is running all the electrical devices on the car—such as

Starting and ignition circuits. (Pontiac)

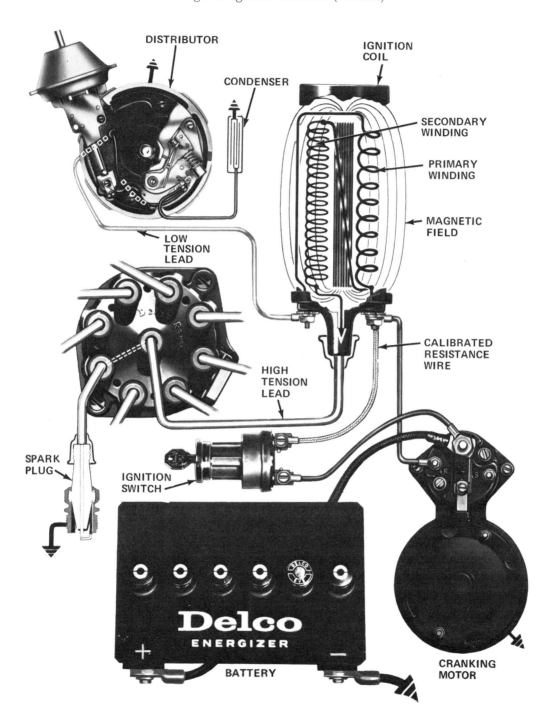

DISTRIBUTOR

CONDENSER

IGNITION COIL

SECONDARY WINDING

PRIMARY WINDING

MAGNETIC FIELD

LOW TENSION LEAD

CALIBRATED RESISTANCE WIRE

HIGH TENSION LEAD

SPARK PLUG

IGNITION SWITCH

Delco
ENERGIZER

+ —

BATTERY

CRANKING MOTOR

lights, horn, and radio—draw current directly from the alternator with the battery being cut out of the circuit.

Since the voltage generated by an alternator depends on the speed of the engine, there must be some device to see that the voltage does not get high enough to burn out bulbs and other electric components. This is accomplished by the *voltage regulator* which resembles a little black box. On a few cars it is integral with the alternator. Voltage regulators used with generators serve more functions than those used on alternator systems.

Inspect the fan belt which runs the alternator. If the belt is loose or frayed, it must be tightened or replaced.

Check the belt for tightness by hooking a finger around it at mid-point and trying to deflect it. A fan belt which is too tight will not deflect at all, even under considerable pressure; while a belt which is too loose will move an inch or more. There are special spring-operated gauges to set the belt to the manufacturer's specifications; however you can get along without one. A belt tightened correctly should move about one-half inch at mid-point under finger pressure.

To tighten or loosen a fan belt you must loosen the bolts which hold the alternator (or generator) on a bracket and pry the alternator to a position where the belt has just the right amount of give. Pry carefully with a wooden board and maintain tension while you tighten the bracket bolt. To replace a fan belt simply loosen it to a point where it comes off easily, then put the new one on and adjust tension. Sometimes there are two belts and the inner one runs the water pump.

Fan belts which are too tight can cause an alternator bearing failure; those which are too loose will slip and the alternator will not put out enough current.

Never pry a fan belt on with a screwdriver; this weakens the belt. The fan belts on air-cooled engines require a different procedure for tightening. See the next chapter for details.

THE FUEL SYSTEM

The fuel supply system is pretty straightforward. There is a gas tank, a pump to bring the gas up to the carburetor, the carb itself which mixes gas

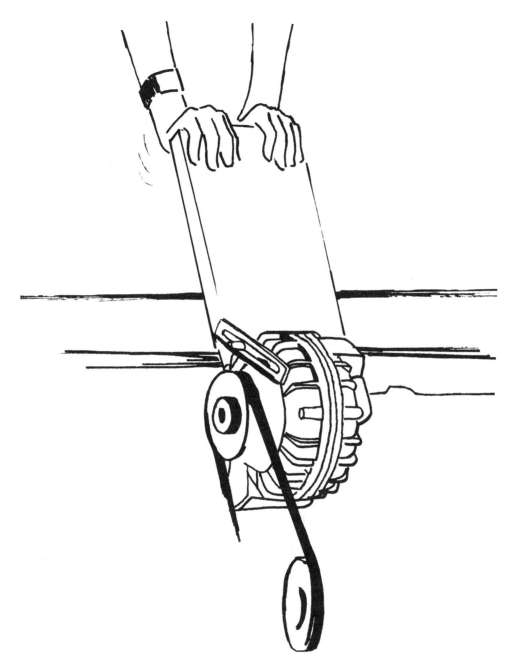

Prying alternator on bracket to remove fan belt.

and air in the right proportions, and an intake manifold through which the mixture flows to each cylinder.

Have someone step on the gas pedal as you look under the hood to see the accelerator linkage move. The linkage is attached to the carburetor. There may be one or more carburetors on your car.

Note the fuel line (either a steel line or flexible tubing) running to the intake on the carburetor. Trace the fuel line back and you should find a bell-shaped pump attached to the side of the engine. This is the *fuel pump* which works by having a diaphragm activated by engine vacuum.

If you do not find a fuel pump on the side of the engine, your car has an electric pump rather than a vacuum-operated one. Electric pumps are located in or near the gas tank since they push the gas up rather than pulling it to the carburetor.

On top of the carburetor there is an *air filter*, which is usually enclosed in a metal canister.

When neglected long enough an air filter clogs up with dirt. It does not allow enough air to pass and "strangles" the engine. The result is stalling, backfiring, poor gas mileage—even unburned gas getting into the crankcase and diluting the oil so that the bearings and cylinders are not being lubricated properly.

Mechanical fuel pump.

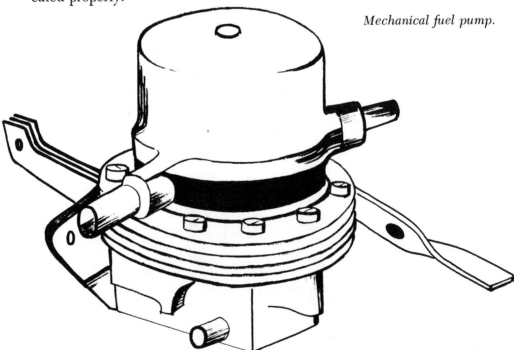

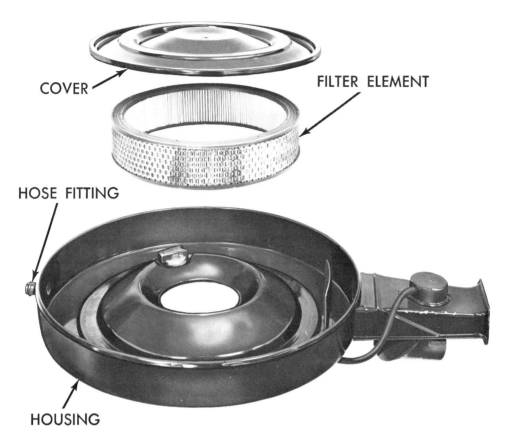

COVER

FILTER ELEMENT

HOSE FITTING

HOUSING

Air filter canister and element. (Dodge)

Changing the air filter is a two-minute job, especially if you have one of the widely used pleated element types. These are made of paper and all you have to do to replace them is open the canister (by unscrewing a wing nut or a couple of bolts) and put the new element in. No tools necessary. On cars with emission controls, hoses going to the filter canister must be pulled off and then replaced.

Replace the air filter at intervals suggested in your owner's manual or at least every 12,000 miles. Buy the new filter at the auto section of a department store and save money. The box the filter comes in will indicate which cars it fits.

In between replacement intervals check and clean the air filter periodically. If it looks visibly dirty, take it down to the gas station and blow it out with an air hose. Always blow from the inside out and keep the hose nozzle a few inches away from the element.

Some air filters are made of polyurethane foam. These must be removed from the holder, soaked in a solvent such as kerosene to clean, squeezed out and dried, then wetted with light motor oil. They are less efficient than pleated paper filters so they should be cleaned frequently. Polyurethane filters can usually be replaced with pleated paper types which fit in the same holder.

Another type of air filter is the oil bath variety. This contains a wire mesh element and separate oil reservoir. To clean this type of filter, take it out and dump the oil. Clean the filter holder in solvent, and wipe any dirt from the holes in the mesh element canister. Then replace and refill with motor oil. Use light oil in the winter and heavier oil during the summer months.

Some older types of filters have a mesh element without an oil reservoir. They are simply cleaned and wetted in light oil before replacing.

COOLING SYSTEM

The cooling system circulates fluid (water plus anti-freeze) around the engine cylinders to keep them from overheating. The fluid is cooled in a radiator where it gets maximum effect from fresh air coming in under pressure. Then it is pumped to the engine by a belt-driven *water pump*. A fan to cool the radiator is also driven by a belt off the crankshaft pulley.

The flow of water circulating to the engine is controlled by a *thermostat* located at one end of the upper radiator hose. Most thermostats work on a bi-metallic principle in which two metal strips with a different expansion rate are bonded together to make a spring. When the engine reaches a certain heat level, the spring lengthens and opens a plate in the thermostat, allowing water to circulate to the engine. Most thermostats are set to open at about 180° to 200° F.

Without a thermostat a car would take a lot longer to heat up. This happens when the thermostat goes bad and remains open at all times. The engine heats up slowly and may stall out until warm. The heater puts out no heat until the engine is finally warm.

A thermostat which fails in the closed position is even more apparent. The engine soon overheats and will not stop doing so until the thermostat is replaced.

All modern cars have a pressurized cooling system. The air pressure (as much as 16 pounds per square inch) causes the coolant to boil at a higher temperature. Each pound of air pressure raises the boiling point about 3° F. Thus the coolant may normally be heated to a range well above the boiling point of water (212° F.) and may not boil until it reaches 250° or more. Newer cars are designed to run hot since this cuts down on the pollutants emitted.

The radiator cap is designed to contain this pressure and has a relief valve which opens when the pressure gets too great. Cooling problems can be caused by a weak or broken spring. Since the cap must be tested by a pressure tester which costs more money than most home mechanics can afford to invest, it is a good idea to carry a spare radiator cap and replace the old one whenever it is suspect.

Pressure-retaining radiator cap. (Dodge)

When the coolant level in the radiator is down, don't add pure water to it. Water is okay in emergencies, but the modern cooling system is designed to run on a water/anti-freeze mixture. For all-year-round protection a 50 per cent anti-freeze mix is fine in cold climates while a 25 per cent mixture will do in warmer zones. In areas where the temperature gets down to below minus 30°, the coolant should be 60 per cent anti-freeze.

Make it a point to use a name-brand ethylene glycol anti-freeze which contains rust inhibitors, water pump lubricant, and an anti-leak compound. Keep some on hand mixed with water in the proper proportions. When the level in the radiator goes down, fill it to about an inch below the overflow tube. Add coolant when the engine is cold.

Ethylene glycol anti-freeze has a higher boiling point than water and safeguards against overheating in hot weather. The same coolant can be kept in the radiator all year and will even last for two years if you add a can of rust inhibitor and water pump lubricant (a canned additive) before the winter of the second year.

Whenever the fluid in the radiator looks rusty, the system should be flushed. The rust does not generally come from the radiator since it is made of brass, or possibly aluminum. Rust is picked up in the engine passages. A procedure for chemically flushing the radiator is covered in Chapter 2.

CLUTCH, TRANSMISSION, AND DRIVE TRAIN

On manual transmission cars the clutch is employed to transmit the engine's power to the gearbox. Logically it sits between the rear of the engine and front of the gearbox.

The engine's crankshaft turns a heavy wheel known as the *flywheel*. The clutch has a round disc lined with friction-producing material which is pushed into contact with the flywheel whenever the clutch pedal is out. When you push the clutch pedal in, the disc moves away from the flywheel. While the clutch disc is in contact with the flywheel, the crankshaft motion is transmitted back to the gears in the gearbox.

In a manual transmission the clutch is mated to a transmission input shaft which has a gear wheel on it. The gear wheel is in mesh with another gear in the transmission—even when the gearbox is in neutral.

A shaft in the transmission (called the countershaft) is being turned by the gear on the input shaft. The countershaft has a series of gears on it and all of these are in constant mesh with gears on another shaft (the output shaft) which runs from the transmission to the driveshaft.

The reason the output shaft is not rotating when the gearbox is in neutral is that all its gears spin freely around the shaft. Now put the car in gear and one gear wheel is locked to the output shaft by a little clutch-like mechanism (called a dog-clutch) which slides over against the gear and has

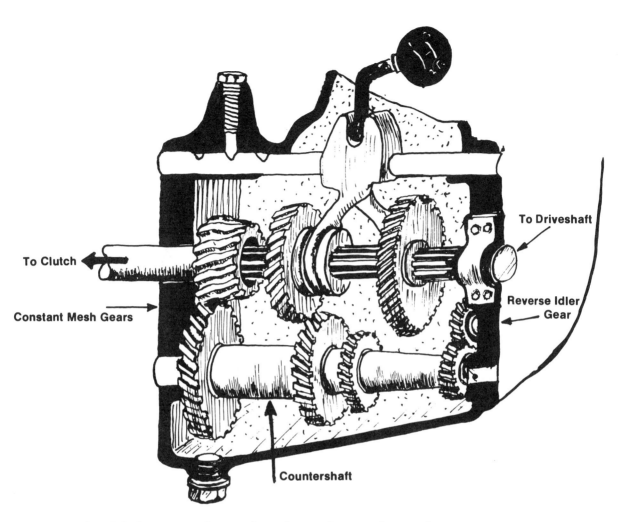

To Driveshaft

To Clutch

Constant Mesh Gears

Reverse Idler Gear

Countershaft

Simplified cutaway of manual gearbox with unsynchronized gears.

teeth that mesh with a set of teeth on the side of the output gear. Each position of the gearshift lever locks up a different gear.

What is the purpose of having all these gears, you may well ask? The answer is that they vary the number of revolutions the engine has to make before it turns the rear wheels once. This is because the engine only develops its full power when it revolves at near-maximum speed. At lower speeds it needs help.

The gearbox gives the engine a mechanical advantage. If the power from a single revolution of the engine can all be applied to turning the rear wheels one-quarter turn, then it can do this four times more powerfully than would be the case if it had to revolve the wheels a full turn.

In first gear the engine can apply a great deal of power to the rear wheels, but it has to revolve very fast to achieve this power (four times as fast as the wheels in our theoretical system). The speed of the car is limited because the engine can turn only so fast and no faster.

Now we progress to second gear. The ratio between the gear wheel attached to the transmission and the one on the output shaft is different. Now the engine has to only turn twice to rotate the drive wheels a full turn. There is less power but a higher top speed.

There are three or four forward gears in most transmissions. Top gear is usually a direct (one to one) ratio. This would mean that the drive wheels would be turned once for every revolution of the engine, except that a further set of gears in the rear axle increases the ratio to somewhere around three or four to one. In top gear the engine revolves three or four times to make the tires rotate once. In low gear the engine may turn fifteen times or more to make the tires turn once.

How does reverse gear work? Simply by interposing another gear between a gear on the transmission countershaft and a gear on the output shaft. This makes the output shaft reverse its direction of rotation.

There is one more clever device in the modern transmission called *synchromesh*. It is needed because the transmission input shaft tends to keep turning due to inertia even when the clutch is not engaged. Thus the transmission gears keep turning a little and they turn the gears on the output shaft which are in mesh with them. If sliding dog clutches alone were used to lock the output gears to their shaft there would be a lot of clashing as they engaged. Synchromesh gears project from the dog clutch and rub on the output gear. They keep the dog clutch moving at the same speed as the output gear so there is no clashing on engagement.

Now what about automatic transmissions? They also change the ratio between the engine's revolutions and the number of turns the rear wheels make, but there is no clutch and no gears to shift.

Without going into the complexities of the modern automatic transmis-

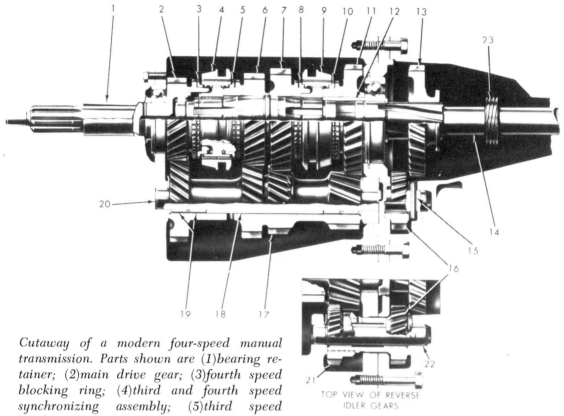

Cutaway of a modern four-speed manual transmission. Parts shown are (1)bearing retainer; (2)main drive gear; (3)fourth speed blocking ring; (4)third and fourth speed synchronizing assembly; (5)third speed blocking ring; (6)third speed gear; (7)second speed gear; (8)second speed blocking ring; (9)first and second speed synchronizing assembly; (10)first speed blocking ring; (11)first speed gear; (12)first speed gear sleeve; (13)reverse gear; (14)main shaft; (15)reverse idler shaft roll pin; (16)reverse idler gear (rear); (17)countergear; (18)countershaft bearing roller space; (19)countershaft needle roller bearing; (20)countershaft; (21)reverse idler gear (front); (22)reverse idler shaft; (23)speedo drive gear. (Pontiac.)

sion, we can understand the basic principle of how power is transmitted and multiplied.

When you see a branch floating in water, you can push your hands in the water and create a turbulence which will move the branch. Now suppose there were two paddle wheels in the water facing each other at a short distance. One paddle wheel is being turned while the other is free-spinning. The turbulence in the water created by the driven wheel will cause the free wheel to spin. In effect, there is a fluid coupling between the wheels.

The automatic transmission has a device known as a torque converter which consists of two turbines set opposite each other in a sealed housing filled with transmission fluid. The turbines are very close but not touching. As one turbine is driven by the engine, it creates a turbulence which causes the other to rotate. Power is transmitted by this fluid coupling.

By putting a device with stationary angled blades around the turbines so that it directs oil flow going off to the side back against the output turbine, the twisting force of the input turbine can actually be increased to as much as two to one. The input turbine is called an impeller and the ring of blades is called a stator. The simple fluid coupling has now become a torque (twisting force) converter because the output force is more than the input force. Some earlier automatic transmissions were composed of a series of torque converters.

The automatic transmissions composed wholly of a series of torque converters proved to be too complex and inefficient, so newer designs use a torque converter to drive a hydraulic pump. Through a maze-like series of valves the hydraulic fluid operates pistons which engage clutches or bands, which then put a variety of gears into mesh. These are called planetary gears since two smaller gears are in constant mesh with a larger one and revolve around it. The gearing is also highly complex.

The load on the engine is sensed by a vacuum-operated valve connected by a hose to the intake manifold. This regulates hydraulic pressure and indirectly affects the speed at which a shift takes place. A governor on the transmission controls the shift points by a series of valves which open to let hydraulic pressure through and cause clutches or bands to activate sets of gears. It all sounds complicated and it is.

After the power gets through the transmission, things are considerably simpler. The power goes from the output shaft through the long drive shaft beneath the car, and the drive shaft has a hypoid gear which allows the power to turn at right angles and drive the gears in the axle shafts going to each wheel. The ratio between the driveshaft gear and the gear in each axle shaft is known as the rear axle ratio. It changes the ratio between the number of turns the engine makes to each turn the rear wheels make in the same way as the transmission does. Unlike the transmission, however, the

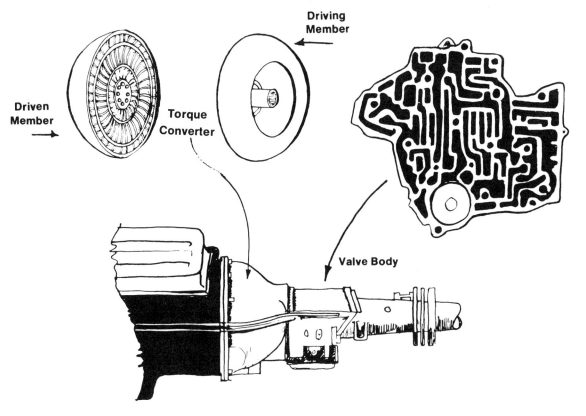

Driving
Member

Driven
Member

Torque
Converter

Valve Body

Components of automatic transmission.

rear axle ratio is fixed and doesn't change unless new gears are installed.

The driveshaft must flex up and down as the car wheels hit a bump or rut and the transmission and one axle remain stationary. To give the driveshaft this flexibility there are U-joints at either end. These U-joints allow the driveshaft to flex yet still rotate and transmit power.

In the rear axle assembly there are differential gears as well as the axle gearing. The differential gears are needed because the inside wheel travels a shorter distance than the outside wheel when the car turns a corner. Without a differential the outside wheel would have to be dragged along to match the speed of the inside wheel. This would make the car hard to control and cause excessive tire wear. The differential gears allow one wheel to slip and match the speed of the other.

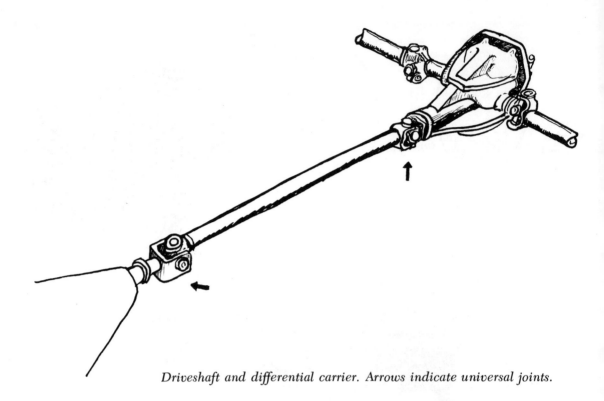

Driveshaft and differential carrier. Arrows indicate universal joints.

The differential can cause problems on ice or snow because power tends to follow the path of least resistance. The wheel which has no traction gets all the power and spins madly while the other doesn't move at all.

Some cars have a limited-slip differential to compensate for this. Each wheel is allowed only a limited amount of slippage due to differential action. When the limit is reached, clutches engage and deliver power to the slipping wheel.

Not all cars have the same type of transmissions and drive-train arrangement. The Swedish Saabs, some models of Renault, Fiat and BMC cars (Austin and Morris), and the Oldsmobile Toronado and Cadillac Eldorado drive the front wheels instead of the rear wheels. VW, Porsche, Corvair, and some Renaults are among the cars which have a rear engine and a transmission in unit with the axle. Some Lancias and the more familiar Pontiac Tempests had a front engine driving the rear wheels but used a transaxle (transmission in unit with the axle) at the rear. Jeeps and similar off-road vehicles drive all four wheels for maximum traction through snow and mud. Each design has its advantages and drawbacks.

You can check the fluid level in the automatic transmission by locating the dipstick near the firewall, to one side of the engine. The engine should be running and at normal operating temperature when you make the check. The fluid is colored red or green and may be difficult to see on the dipstick. Hold it against a white sheet of paper and check carefully. Each mark below the top one usually indicates that the fluid is down one pint.

Ford products use Type F transmission fluid while GM and Chrysler cars require Dexron. The fluid is poured in through the dipstick tube so you will need a funnel.

Transmission fluid which has turned brown or yellow, or which has a burnt smell, is an indicator of transmission problems. A low fluid level means leakage since the fluid is not normally used up. Consult a mechanic when you suspect that the transmission is not operating normally.

If you have power steering on your car, the fluid level is also checked with a dipstick. The power steering mechanism consists of a belt-driven pump and a fluid reservoir. Locate the pump near the front of the engine and check the dipstick with the engine off. When the level is down you can add the same fluid as you use in your automatic transmission. However, some manufacturers recommend a special power steering fluid which works a bit better.

Ring and pinion gears in rear axle conduct power to wheels by allowing power to turn at right angles (as indicated by arrows).

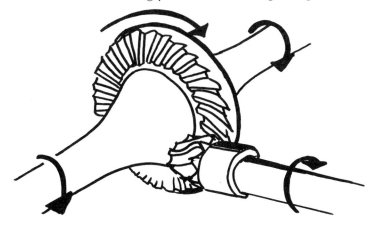

THE BRAKING SYSTEM

When you push down on the brake pedal, semi-circular "brake shoes" covered with material of special composition are forced outward against a steel brake "drum" inside each wheel and the car stops. How does this happen?

At one time, translating your force against the pedal to a force against the brake drums was done by mechanical linkage. The trouble is that the force against all wheels must be in a definite proportion and mechanical brakes were always going out of adjustment and losing this proportion.

The solution was the modern hydraulic brake system. It is based on the simple physical principle that force applied on a fluid will be transmitted equally in all directions by that fluid.

Moving the brake pedal pushes a piston in the *brake master cylinder*. The piston pushes on the brake fluid. The brake fluid carries the push

Rear brake assembly. (Ford)

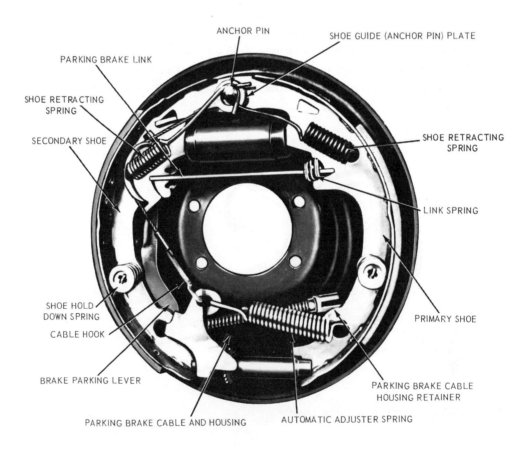

ANCHOR PIN

SHOE GUIDE (ANCHOR PIN) PLATE

PARKING BRAKE LINK

SHOE RETRACTING SPRING

SECONDARY SHOE

SHOE RETRACTING SPRING

LINK SPRING

SHOE HOLD DOWN SPRING

CABLE HOOK

PRIMARY SHOE

BRAKE PARKING LEVER

PARKING BRAKE CABLE HOUSING RETAINER

PARKING BRAKE CABLE AND HOUSING

AUTOMATIC ADJUSTER SPRING

through the brake lines to a cylinder in each wheel known appropriately as the "wheel cylinder."

The fluid "push" in the wheel cylinders forces apart two rubber-capped pistons which move against the brake shoes and shove them into contact with the brake drum.

When you take your foot off the brake pedal, a strong return spring moves the brake shoes away from the drum, back to their original position. This also produces a fluid force which is transmitted from the wheel cylinder back to the master cylinder and returns to the holding reservoir the small amount of fluid displaced. The cycle is complete.

Brakes need service when the pedal engages at a lower point than usual or feels spongy. A simple test for leaks in the hydraulic system is to hold the brake pedal down firmly for at least a minute. If the pedal slowly sinks to the floor, there is a leak somewhere which is allowing fluid to escape.

Locate the brake master cylinder on your car. Most master cylinders are in a prominent position in the engine compartment, on the firewall. On a few cars the master cylinder is hidden away beneath the floor boards or in some other out-of-the-way location.

To check the brake fluid level, just unscrew or unbolt the master cylinder cap and look in. The level should be one-quarter inch from the top.

When replacing brake fluid, look for a can labeled as meeting specification 70R3 or J1703A. This fluid has a boiling point of about 375° and offers the best protection for drum brakes.

Master cylinder reservoirs should be filled to within ¼ inch from top, as indicated in diagram. (Chevrolet)

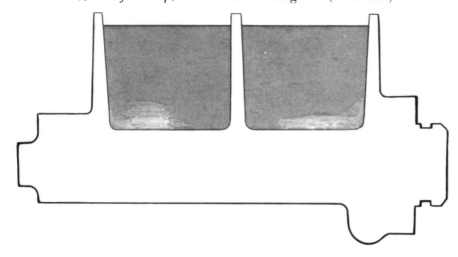

Disc brakes require a special fluid with a higher boiling point. No specification has been established as yet, so buy a fluid specifically labeled as being for disc brakes. The best ones have boiling points as high as 550°.

Be sure to replace the master cylinder cover and bolt it down tightly.

INSPECTING TIRES

Look at each tire closely for evidence of lumps, excessive cracking on the sidewalls, cuts, or baldness. Keeping good rubber on the car is an essential safety measure.

The traditional way to check for wear is to stick a Lincoln-head penny between rows of tread that seem worn down. If the penny is covered from its edge to the top of Mr. Lincoln's head, then you have one-sixteenth of an inch of tread and the tire is safe (and legal). When the tread is less than one-sixteenth of an inch across the tire, it must be replaced.

Tires worn unevenly are an indication of defective shocks, out-of-balance wheels, habitual over or under inflation of tires, or wheels which are out of alignment. You can check the shocks and tire pressure easily, while you may wish to leave wheel balance and alignment to a garage which specializes in this kind of work.

All new tires have inflation recommendations marked on the sidewall of the tire. For better handling, at the expense of a slightly stiffer ride, many drivers like to inflate tires about five pounds over recommended pressures all around.

You should have your own air pressure gauge since the ones on air hoses at gas stations take a lot of abuse and will frequently give an incorrect reading. Dial-type pressure gauges are easiest to read, but the pencil style gauges are cheaper and less fragile.

On some cars it may be necessary to remove a hubcap in order to get the pressure gauge firmly seated on the tire valve. You can pry the hubcap off with a large screwdriver or the lever on one end of some lug wrenches. Replace the hubcap by hitting it lightly with the heel of your hand all around. Be sure it snaps in firmly on all sides.

To avoid the trouble of removing hubcaps in order to check the air pressure, buy valve stem extenders at an auto accessory store. The cap on the end of the valve stem is also an inexpensive accessory store item. Always

More recent tires are manufactured with indicators or "wear bands"
which show by break in tread pattern that tire should be replaced.
In diagram on left, tread is still good; on right, it is worn out.
(Chevrolet)

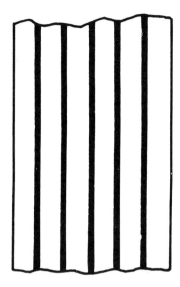 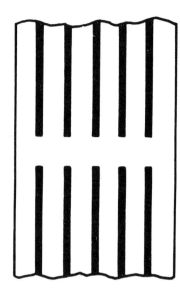

replace a missing one since it provides extra protection against losing air.

Make a note of which tires are down in pressure so you can pump them up when you go to get gas.

TESTING SHOCKS

Worn or defective shocks will cause a car to rebound loosely when going over bumps and may lead to loss of control. Thumping and thudding noises from the chassis when going over ruts or bumps are another sign of failing shocks.

Some people prefer to check shocks while the car is stationary. The procedure is to push down on each bumper with your foot until the car is rocking on its springs. When you take your foot off the bumper, the car should make a single cycle before returning to level position. Any extra jiggle indicates that the shocks are worn.

Original equipment shocks on most cars are marginal. They may need replacing after as little as 20,000 miles. The front shocks usually fail first since they carry the most weight on a front-engined car. The replacements should be heavy-duty shocks which cost $10 apiece and up. These shocks will snub the springs more efficiently and may seem harder-riding to some. However, they will last much longer and improve a car's handling significantly.

On station wagons, or other cars which consistently carry heavy loads, shocks with auxiliary springs around them may be fitted. When used with adequate tires, they can add as much as 500 pounds to the load-carrying capacity of a car.

Front and rear shocks should usually be replaced in pairs. When one is failing, the other is likely to have a limited life expectancy. The exception is when one shock begins leaking prematurely. In that case it's all right to replace the single faulty shock.

ELEVATING THE CAR

To inspect shocks closely for fluid leakage (which usually can be seen as a dark stain on the dirt covering the shock body) it may be necessary to raise the car while you crawl under it.

To elevate the car you will need a jack plus safety stands or blocks to place under the suspension arms or axle. Working under a car supported by a jack alone is unsafe.

Jack safety stands cost about $6 apiece and they are a good investment. Heavy boards can also be used to support the car if you do not have stands. Cement blocks are not as safe since they can crumble. If you must use cement blocks, be sure they are new ones and use two sets on each side to build up firm support.

Try to select a level spot for jacking up the car so that it won't roll. Use wedge-shaped chocks under the opposite wheels as an extra precaution. If you don't have any wheel chocks, use small stones. You can even use the spare tire in front of one wheel to keep the car from rolling. Be sure to set the handbrake and put the car in gear or in the park position on an automatic transmission.

Bumper jacks are easy to place. Just be sure they are near the end of the bumper on the side you are jacking and that the raising "lip" of the jack is

Wheel diagonally opposite to the one being elevated should be firmly chocked as shown. (Volkswagen)

firmly wedged under the bumper. Many cars have bumper slots to fit the jack lip.

If you are using the bumper jack that came with the car and have never used it before, it may be a bit confusing. The base, upright, and handle are separate. The upright is inserted into the base, then the handle (which is also the lug wrench) is fitted into the ratcheting mechanism of the upright. A switch (metal-finger type) controls whether the mechanism ratchets upward (raising the car) or downward when you pump the handle.

Scissor jacks or hydraulic jacks which go under the car require some care in placement. Jacking in the wrong place might cause you to bend a steering arm or some sheet metal. To jack a front wheel, place the jack under the lower front suspension arm as close to the wheel as possible. For lifting a rear wheel the jack should be placed under the axle where the springs attach. Always look before you jack to get the right spot.

Some foreign cars such as VW and Mercedes have jacks which fit in a special socket under the side of the car.

Raise the car about one foot. Place the jack safety stand under the lower front suspension arm or rear axle (depending upon which end you are raising), then extend the stand as much as you can, and set the pin. Lower the jack so that the car is resting solidly on the stand. Do this on both sides of the car. Test the support by trying to rock the car sideways. It should not move.

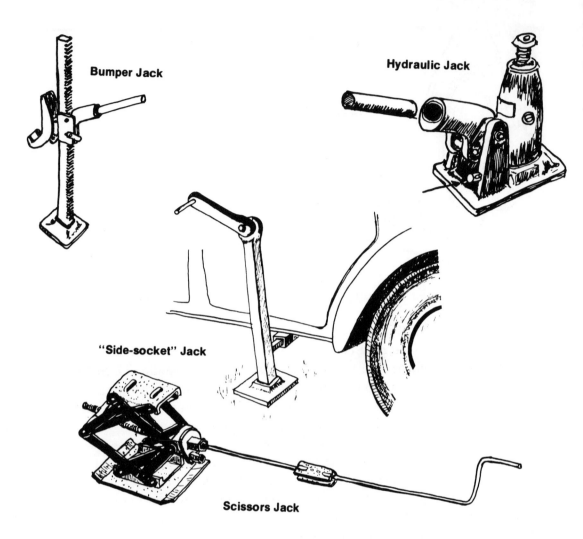

Bumper Jack

Hydraulic Jack

"Side-socket" Jack

Scissors Jack

Types of jacks.

Use an old shower curtain or plastic tablecloth spread on the ground to keep from getting dirty as you crawl under the car. Inspect all shock absorbers for visible signs of leakage. Most shocks cannot be repaired so a leaky shock must be replaced. The expensive adjustable shock absorbers do come apart and new fluid or a piston can be put in.

CHANGING TIRES

Many car manufacturers recommend tire rotation at 5,000-mile intervals. While your car is up on safety stands it is convenient to switch tires. The spare goes on the right rear wheel and all other tires are switched diagonally. The right front tire becomes the new spare.

Unless you have four safety stands or enough blocks to elevate the front and rear of the car, you will not be able to switch all tires at once. On the end of the car which is not elevated, replace one tire at a time.

Whenever it is necessary to change a tire, the general procedure is to begin by jacking just a little to put a load on the jack but not so much that the wheel is off the ground. Then pry off the hub cap and unscrew the lug nuts.

Lug nuts loosen counter-clockwise on most cars. On Chrysler products before 1971 the lugs nuts on the driver's side turn clockwise to loosen while those on the other side go counter-clockwise.

Safety stands must be used when working under car.

Lug nuts which have been tightened by a garage man with a pneumatic wrench can be stubborn. A good cross-type lug wrench should provide plenty of leverage and you can stomp on it if you have heavy shoes or tap it with a hammer. For maximum leverage slip a piece of hollow plumbing pipe ("cheater bar") over the end of the lug wrench. A sharp tug will give more force than a steady pull. Penetrating oil can help free up a corroded lug nut.

Loosen all lug nuts with the wrench. Then jack the car up so the wheel is just off the ground. Unscrew the lug nuts the rest of the way by hand and take the tire off the wheel by pulling it up and toward you.

Rotate the new tire so the bolt holes line up with the lug bolts and ease it on. When screwing on the lug nuts, be sure the narrow end of each is facing the car. This is the only way the nuts will be really tight.

Screw the lugs hand tight only. Then lower the car so that the tire touches the ground before you complete the tightening. Do not tighten one lug nut all the way and then go on to the next. Instead tighten each nut only fairly tight and do diagonally opposite nuts in pairs. This will keep the wheel from warping by being tightened excessively on one side and not at all on the other. Complete the final tightening by diagonal pairs again. Use considerable force on the lug wrench but do not use a device such as a cheater bar since you would be in danger of overtightening the nuts.

THE EXHAUST SYSTEM

Inspect the exhaust system while the car is still elevated. There is an exhaust manifold carrying the exhaust gases from each cylinder, single or dual exhaust pipes connecting to the manifold, and a muffler and tailpipe at the end of each exhaust pipe. Some cars have a small muffler, called a resonator, behind the main one.

Exhaust pipes corrode externally because of road deposits and salts. They also corrode from the inside due to condensation of gaseous exhaust products. Look for signs of rusting through, and tap the pipes and muffler with a screwdriver handle to test for weak spots. A dull "clunk" instead of a metallic "clank" is an indication that the rust goes all the way through. Exhaust repairs are covered in Chapter 3.

Check the brackets that hold the exhaust pipe on. They may also be badly corroded and rusting through. New brackets can be purchased at an auto parts house and bolted on. As a temporary measure you can use wire to lash the exhaust pipe tight.

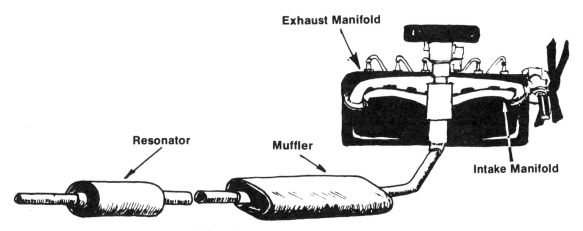

Manifold and exhaust system.

By this point you should be acquainted with the major systems of a car and have identified all the components on your car. In the next four chapters we will cover routine maintenance and repairs you can do yourself.

2

Lube and Cooling

● Before going on to do any serious work on your car you will need some tools. Nothing exotic is necessary; a set of combination wrenches, a few pairs of pliers, a socket wrench set with a ratchet handle, and regular and Phillips screwdrivers in three sizes will do.

Avoid cheap tools because (a) they don't function as well and (b) they break. This is true even of an ordinary screwdriver. A cheap one has a handle which is not shaped for maximum leverage, chrome plating that flakes off, and poor bonding between the handle and shank.

Professional auto parts houses carry top-quality tools, but these are quite expensive and probably better made than they need to be for your purposes. The top line of tools at a department store or hardware store is almost as good and a lot more reasonable. Brand names such as Sears's Craftsman, Ward's Powr-Kraft, Mac, Husky, S-K Wayne, and Crescent are all fine buys. (See Appendix C for mail order addresses.)

Buying tools in a set can save money. The combination wrenches should be sized from ⅜ inches to ⅞ inches by sixteenths, or a set of at least nine wrenches. A basic socket set consists of sockets in these same sizes, a ⅜-inch drive ratchet handle which fits the sockets (the drive mechanism is a ⅜-inch square which snaps into the square opening in each socket), and at least one short extension piece. There should also be a special ¹³⁄₁₆-inch deep socket with a neoprene insert for spark plugs.

The most versatile pliers for automotive use are probably water pump or Channelock pliers which have parallel jaws opening to various widths. A pair of pliers which crimps solderless terminals and cuts wire is also a good investment.

For lubricating your car you will need a few special tools. The prime requirement is a grease gun which accepts standard-size grease cartridges. Buy a flexible adapter for it along with regular and needle-nose tips.

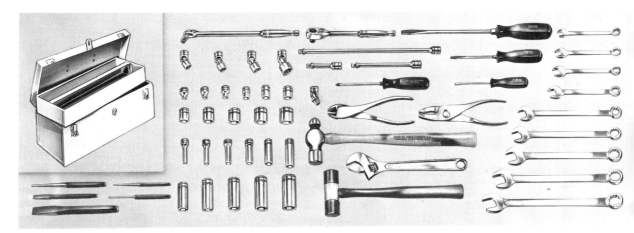

Some basic tools for the aspiring mechanic include punches, socket set adapter handles, screwdrivers and pliers, ballpeen and soft-faced hammers, and set of combination wrenches. (Snap-On)

A funnel which ends in a long flexible tube will allow you to pour in oil almost anywhere. To drain or refill the grease in gearboxes and differentials you will need a gear lube syringe along with a trans-diff wrench to open drain and filler plugs. The latter has square projections of various sizes on either end and fits depressions in the plug caps.

CHANGING YOUR OIL AND FILTER

There is no job which will save you more money for as little effort expended as an oil change you do yourself. It takes only a few minutes and you save by buying oil in bulk as well as by avoiding a labor charge.

Changing your oil and filter at least as often as the owner's manual recommends will keep your car's engine operating many miles longer before major repairs are necessary.

In the new system for grading oil, the highest-quality oils are marked "SD" and "SE" on the can. SE oil is the best and it is designed for powerful cars driven hard. SD oil is for cars which are not used for heavy hauling or driven at prolonged high speeds.

Some oils may still be marked with the old grading system. Prior to 1968 the best oil was marked "MS." Then further requirements for oil were published by car manufacturers and a higher-grade "MS" oil carried the statement that it had passed the car manufacturers' sequence tests. This oil is equivalent to the newer SD grade.

The weight or viscosity of an oil is also a consideration. Heavier oils, such as 30 and 40 weight, are best for use in hot weather or for high-speed driving after the engine is warm. Lighter oils, such as grades 5, 10, and 20, retain their free-flowing properties better in cold weather. A multi-viscosity oil—10W-30 or 10W-40—is good all-year-round under most driving conditions. Where the weather is consistently below minus 10°, a 5W-30 oil should be used. Single weight oils are less expensive and almost as good. Their sole failing is that they will not lubricate as well under conditions where the engine goes from very cold to quite hot, such as after high-speed travel in cold weather. Always use detergent oil and buy your oil in bulk to save money.

You may not have to put your car up on stands to drain the oil. On most cars you can reach the oil pan drain plug with a wrench by just crawling under a bit.

First drive the car around or let it idle until it reaches normal operating temperature. This will heat the oil and allow it to drain freely.

Use a shallow basin, an old roasting pan, or even a five-gallon can with the side cut off as a container for the old oil. Look for the square oil pan under the car. It should be below the engine, possibly somewhat to the rear. Feel around for the drain plug on the pan and put your container under it. Loosen the drain plug with the box- or open-end of a large combination wrench. Allow the oil twenty minutes to drain.

While waiting for the oil to drain completely, you can change the oil filter. Most newer cars have "spin-off" filters which look like a tin can with rounded corners. The filter is screwed on a holder on the side of the engine. Some filters are placed so low that you can only reach them from beneath the car.

Spin-off oil filters are widely available from auto accessory stores and they come in a box marked with the make of car and the particular engine they fit. Some older cars use a replaceable-element filter which fits in a metal canister. With these types of filters you only have to purchase a new element.

Put your drain pan beneath the filter before you take if off, since the

Replacing oil filter. Rubber gasket (arrow) on spin-on filter should be coated with engine oil before fitting. (Ford)

filter contains oil. When changing a filter you will have to add one additional quart of oil to take care of the filter capacity.

There are oil filter wrenches which fit around spin-off type filters, but you should be able to find a substitute. Tools used in the kitchen to loosen jar covers will sometimes work on a spin-off filter. A chain wrench, strap wrench, or even an old leather belt can do in a pinch.

A wrench is not used to screw the new filter on. The correct procedure is to moisten the rubber gasket on the bottom of the filter with fresh oil and then screw the filter on as tightly as you can by hand.

The replaceable-element filter is just as easy to change. The canister is opened with a wrench, the old element is discarded and the canister is cleaned with a rag and solvent, then the new element is inserted.

Finish changing the oil by wiping off the threads of the drain plug and the threads into which it is fitted. Then replace the drain plug tightly. Add the new oil up topside and you are done. Run the engine a bit while you check for leaks around the filter and at the pan.

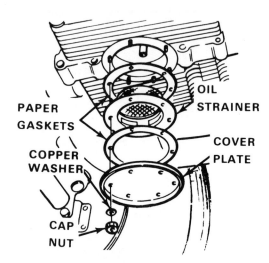

PAPER GASKETS

COPPER WASHER

CAP NUT

OIL STRAINER

COVER PLATE

To drain oil from VW Beetle, parts must be removed in order shown. There is no spin-on filter, so strainer must be cleaned.

After changing your oil you may have a problem disposing of the dirty oil. Pouring it down a drain causes pollution. You can take it to a garage if the owner is a friend of yours since garages store dirty oil in big drums and sell it for reprocessing. Otherwise try bringing it to a local landfill in a container marked "oil." Old oil is sometimes also poured on dirt roads to keep down dust. Don't do this indiscriminately as the slick may cause someone to skid.

GREASE JOB

The days of needing a grease job everytime you change your oil have passed. However, the cars which were supposed to need greasing only at 30,000-mile intervals have not proven as trouble-free as their manufacturers anticipated. Here's where you can profitably deviate from the recommendations in your owner's manual and grease more often than the book says. A thorough job once a year (every 12,000 miles) is advisable.

Older cars had a fair number of chassis points that needed greasing. A grease nipple was fitted at each of these points to accept the nozzle of a grease gun which forced the grease in under pressure. Modern cars have fewer fittings and many of these are sealed with a rubber or metal plug which must be replaced with a nipple fitting or tip adapter whenever you grease.

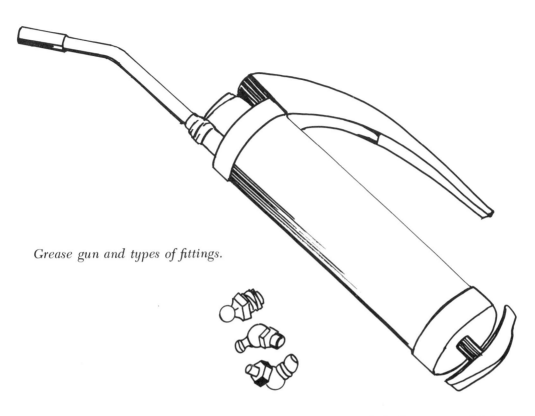

Grease gun and types of fittings.

Service stations use a pneumatic pressure gun and bulk grease which comes in pails. You can make do with an inexpensive hand gun which uses grease cartridges.

The only sure guide to all grease points on a car is the factory service manual or the charts showing greasing points which manufacturers supply to service garages. Ask a friendly mechanic to show you a chart covering your car.

Extended-interval chassis grease should be used (it's marked on the cartridge). Grease in a reasonably warm place since the grease will not flow well at temperatures below freezing.

Wipe each grease nipple and ball detent clean before greasing. Keep pumping grease until the old grease begins to ooze out of the fitting. Fittings with neoprene seals (such as those on sealed ball joints) should be greased only until the seal swells. Pumping in more grease could break the seal.

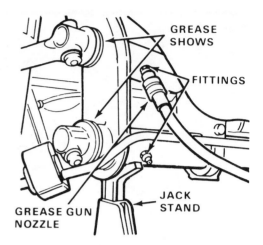

GREASE SHOWS

FITTINGS

GREASE GUN NOZZLE

JACK STAND

Grease until lubricant shows or rubber seals begin to bulge. (Volkswagen)

A needle-nosed grease fitting adapter is required on some U-joints. Another tip adapter can be used with fittings sealed by a plug. Do not permanently install a grease nipple in place of a plug fitting since it won't hold the pressure.

When grease is pumped in, there should be no stress on the part being greased—otherwise your gun won't have the pressure to force enough grease in. Generally you will have taken pressure off the grease points simply by having your car up on jackstands. If you work from a pit underneath the car, you may have to jack the frame up a little to take the stress off.

The differential, manual transmission case and steering box also require lubrication. Sometimes the differential and transmission use the same lube (SAE 90 weight or so) and sometimes not. Limited-slip differentials almost always require a special type of lubricant. Some steering boxes use differential grease while others require regular chassis grease or even motor oil. Always follow the recommendations in your owner's manual to be sure.

Steering boxes may have a grease nipple or a filler plug. The transmission and differential usually have filler plugs with a square depression in them. A trans-diff adapter wrench is studded with squares to fit the depressions in all sizes of plugs. You may be able to open a filler plug with a square bolt held by a pair of vise-grips, but a trans-diff wrench is an inexpensive investment.

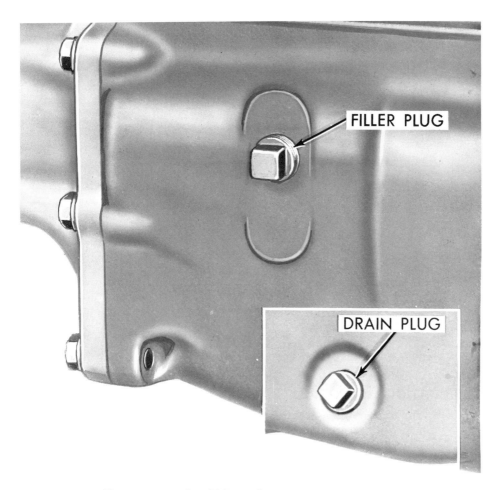

FILLER PLUG

DRAIN PLUG

Transmission should have drain plug beneath hous-
ing as well as level-check and filler plug on side.
(Dodge)

Grease should be up to the level of the filler plugs in the transmission and differential. If not, use a gear lube syringe to fill up to the filler plug level with the recommended lubricant. Replace plugs tightly.

Some manufacturers recommend draining the transmission before putting in fresh lubricant. If this is a routine maintenance procedure on your car there will be a drain plug at the bottom of the transmission case.

Syringe-type pump can be used to fill rear axle or gearbox with grease. (Dodge)

LUBRICATING AND ADJUSTING BODY PARTS

The working parts of an automobile body—latches and locks on doors, hoods, and trunks—will last longer and operate better with a little attention.

Door, hood, and trunk latches and striker plates should be lubricated twice a year with a stainless lube stick applied lightly to the mating surfaces of latches and striker plates. Hinges should be sprayed with aerosol silicone lubricant.

Silicone lube may be sprayed into the window channels of windows that are difficult to crank up and down.

Door or trunk locks that work hard should be lubricated with graphite squeezed into the lock mechanism. Work the key in the lock until it eases up.

When a door remains slightly ajar or requires slamming to close, the striker plate can be adjusted. This is the mechanism which the door latch fits into, and it holds the door with a rotary closer. The striker is adjusted by

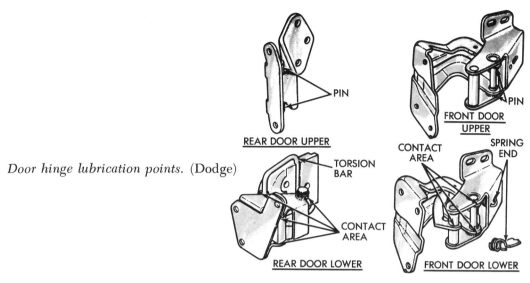

PIN

REAR DOOR UPPER

PIN

FRONT DOOR
UPPER

CONTACT
AREA

SPRING
END

TORSION
BAR

CONTACT
AREA

CONTACT
AREA

Door hinge lubrication points. (Dodge)

REAR DOOR LOWER

FRONT DOOR LOWER

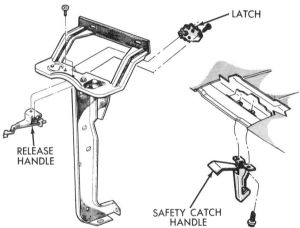

LATCH

RELEASE
HANDLE

SAFETY CATCH
HANDLE

Hood latch and release. (Dodge)

loosening its screws slightly and using a block of wood to tap it inward or outward. Try moving the striker outward slightly and see if the door latches more easily. If not, move the striker inward a bit. Experiment until the door works with ease and latches securely.

The hood latch generally consists of a metal pin in the upper mechanism which fits a hole in the lower latching assembly. If the pin and hole do not line up exactly, the holding assembly can be slightly shifted to either side by loosening its attaching screws a little and tapping it into place with a block of wood.

When a hood won't stay shut, the pin part of the assembly may need adjustment. This is done by loosening the lock nut above the pin and turning the adjusting screw to lower the pin. Each turn of the adjusting screw will lower the pin slightly. Tighten the lock nut after the pin has been adjusted.

The trunk latch generally has a locking mechanism adjusted in the same way as a door striker plate. Just loosen the fasteners and tap with a block of wood until alignment is achieved.

When lubricating and adjusting body parts don't forget to spray lube on the little hinged door over the gas cap, removable fender skirts, and any other moving parts. Spray your jack and lug wrench with silicone also to prevent rusting.

Rubber parts, such as weatherstripping around doors and trunk, need lubrication also. Coat lightly with a special rubber lubricant (Ru-Glyde, for example). To prevent weatherstripping from sticking in cold weather you can use the same anti-freeze used in your radiator. Brake fluid is also an acceptable rubber lubricant.

Trunk latch. Silicone spray may be used on contact surfaces. (Dodge)

LUBRICATION: OPERATING PARTS

Along with a grease job and lubrication of body parts, most cars have other operating parts which must be periodically treated with light machine oil or another recommended lubricant. The parking brake cable, clutch shaft, steering arm stops, shift linkage, and accelerator linkage are among parts

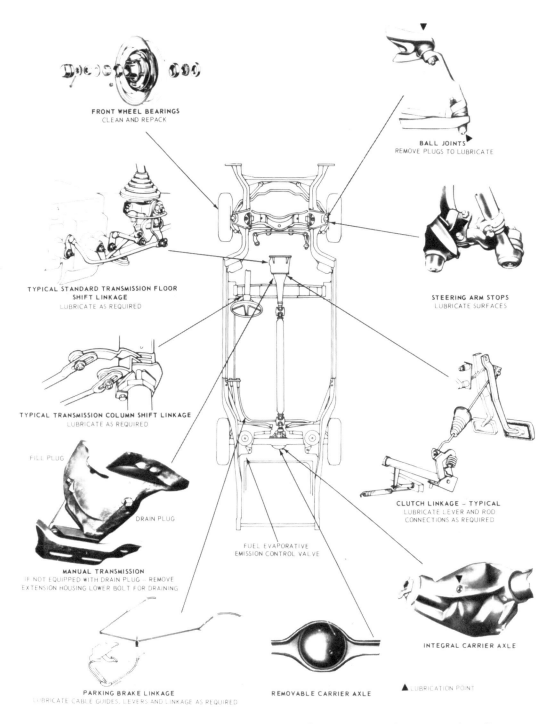

FRONT WHEEL BEARINGS
CLEAN AND REPACK

BALL JOINTS
REMOVE PLUGS TO LUBRICATE

TYPICAL STANDARD TRANSMISSION FLOOR
SHIFT LINKAGE
LUBRICATE AS REQUIRED

STEERING ARM STOPS
LUBRICATE SURFACES

TYPICAL TRANSMISSION COLUMN SHIFT LINKAGE
LUBRICATE AS REQUIRED

FILL PLUG

DRAIN PLUG

CLUTCH LINKAGE – TYPICAL
LUBRICATE LEVER AND ROD
CONNECTIONS AS REQUIRED

MANUAL TRANSMISSION
IF NOT EQUIPPED WITH DRAIN PLUG – REMOVE
EXTENSION HOUSING LOWER BOLT FOR DRAINING

FUEL EVAPORATIVE
EMISSION CONTROL VALVE

INTEGRAL CARRIER AXLE

PARKING BRAKE LINKAGE
LUBRICATE CABLE GUIDES, LEVERS AND LINKAGE AS REQUIRED

REMOVABLE CARRIER AXLE

▲ LUBRICATION POINT

Typical chassis and powertrain components which require periodic service. (Ford) frequently requiring lubrication. Generators on older cars often had an oil cup which required periodic filling. Distributors usually require a few drops of oil on the felt pad in the distributor cam each time the car is tuned. A special lubricant is used on the cam itself. Rubber bushings and grommets on suspension arms and steering require a special lubricant, such as Ru-Glyde, which does not harm the rubber.

Manufacturers' recommendations on lubrication can be found in your owner's manual. Following them religiously will cut down on wear, cut out squeaks and groans from dry parts rubbing against each other, and make your car operate better longer.

GAS TANK AND OIL PAN LEAKS

Rocks can puncture a gas tank, oil pan, or automatic transmission fluid pan and cause leakage. Often the tank or pan will have to be replaced, but you can try small repairs yourself with a kit consisting of epoxy resin and hardener plus fiberglass cloth. Such kits are sold by J. C. Whitney and by many hardware stores.

Clean surfaces around the hole with solvent and dry. Mix up some epoxy and apply it around the hole. Cut a piece of fiberglass cloth to overlap the hole at least an inch on all sides. Fit the cloth over the hole so it is held on by epoxy and cover the cloth with more epoxy mix. For a small hole use two layers of cloth and epoxy. A somewhat larger hole will require three or four. Allow the epoxy to dry for 12 hours before using the car.

When the leak area cannot be kept dry, use a special type of epoxy sold by marine supply houses which can set while constantly dampened.

A missing oil pan drain plug, or one with stripped threads, can be replaced with a universal rubber drain plug made for the purpose. It comes with a tool to insert it and is sold by many car accessory shops.

SPEEDOMETER CABLES

The speedometer in most cars measures the revolutions of a pinion gear in the transmission, translating this into miles per hour. On other cars the speedometer directly measures rotation of one of the wheels.

The speedometer head contains the measuring mechanism and is fairly complex. You should not attempt repairs yourself unless instructions are given in the shop manual and you are confident of doing the job right.

Fortunately speedometer heads are relatively trouble-free. It is the cable which connects the head to a pinion gear or wheel that is most likely to go out of kilter. Often the first indication is a loud tick or grating noises from the vicinity of the speedometer indicating that the cable needs lubrication. Sometimes the speedometer needle fluctuates wildly (indicating a kinked cable or cable housing) or fails altogether (usually a snapped cable).

To get at the cable it must be pulled from the cable housing at the point where it feeds into the head behind the dashboard. Most likely you can reach the attaching point by feeling up under the dash. On some cars you must remove the instrument cluster or another part of the dash to get at the cable.

To detach the cable housing from the head you may have to loosen a nut or else (on many late models) press in a quick disconnect coupling. Either way, it's simple.

If the connecting nut was loose when you took if off, this may have been the problem. Re-tighten the nut and see if the speedometer works.

Crawl under the car and trace the cable to see whether it runs to the transmission or to one of the front wheels. Cables to a wheel must be disconnected at the wheel end also before removal. Take off the hub cap and note where the cable attaches to the grease cover (small inner hub) over the wheel bearings. The cable is disconnected by pulling a cotter pin with a pair of pliers.

Slowly pull the cable from its housing. Any hang-ups indicate that the cable and housing are both bent and must be replaced with new parts obtained at an auto parts store.

With the cable out of the housing, inspect it for kinks or bends, especially at the tip. Roll the cable on a flat surface to be sure there are no kinks not readily visible.

Lubricate the cable with special speedometer grease or Lubriplate. Apply it sparingly and do not grease the cable tip (which goes into the head) at all.

Feed the cable back into the housing with care to prevent kinking. Then reconnect everything. Speedometer cables which were disconnected from a wheel will require a new cotter pin.

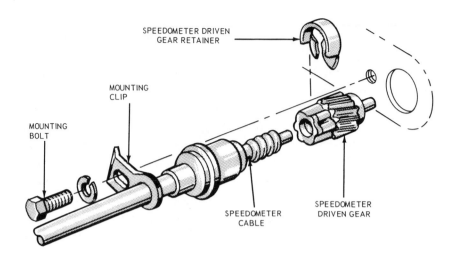

SPEEDOMETER DRIVEN
GEAR RETAINER

MOUNTING
CLIP

MOUNTING
BOLT

SPEEDOMETER
CABLE

SPEEDOMETER
DRIVEN GEAR

Speedometer cable assembly is usually geared to the transmission. (Ford)

A snapped speedometer cable must be replaced. Only one part can be pulled out from the head end, the other must be removed from the wheel or transmission end.

Cables going to the transmission pinion gear are just as easy to disconnect as those going to a wheel. Just use a wrench to loosen the coupling. When the broken part of the cable is out, reconnect the coupling and feed the new cable in from the head end. You cannot remove an unbroken cable from the transmission end because it has a limiter stop-nut.

One symptom of a faulty speedometer is usually caused by the head itself. This is when the needle will not go past a certain point on the dial. Before going to a professional for repairs, however, check and see whether the needle itself is bent.

Speedometer cables should be lubricated once a year as a matter of routine maintenance. Lubrication is also necessary whenever a cable becomes noisy. Otherwise the cable is likely to snap.

LUBRICANTS, SOLVENTS, AND ADDITIVES

Quite a few lubricants are required to do a thorough job of greasing and oiling your car. Along with motor oil and light machine oil, there is chassis and differential grease, wheel bearing grease, silicone spray, graphite for locks, stainless stick lubricant for doors, penetrating oil (such as Liquid Wrench) for freeing rusted nuts and bolts, distributor cam grease, speedometer cable grease, and rubber lube.

Other maintenance procedures will require aerosol solvent for automatic chokes and exhaust manifold heat valves, a parts-cleaning solvent such as kerosene or Mr. Clean, and engine degreaser.

As long as you use good motor oil, additives are not necessary. Oil additives are big business—in some cases multi-million dollar business that sponsors races, gets endorsed by sports figures, and advertises in the slick magazines—yet, despite this high-pressure salesmanship, there seems to be no definitive proof that they are a bit more effective than oil alone, provided that it is the proper grade of oil.

Super coolants for radiators, magic stuff to extend battery life, and cure-all fluids for automatic transmissions are similarly dubious. To my knowledge there has been no test by an unbiased source to prove that any of them do the least bit of good.

COOLING SYSTEM

The only special tool required for routine cooling system service is a radiator hydrometer.

Ethylene glycol anti-freeze is heavier than water, so the percentage of anti-freeze in the radiator coolant can be determined by the amount of coolant a float inside the hydrometer displaces. More expensive hydrometers have a scale showing the degree of temperature the mixture will go to without freezing. Cheaper ones have floating colored balls to give an indication of the temperature range.

Test the anti-freeze by drawing some coolant into the hydrometer (it looks like a giant eye-dropper and works on the same principle) and holding it level while you read the float height.

RADIATOR AND HEATER HOSES

Rubber hoses are used to convey radiator coolant to the engine block and back again to the radiator. Narrower hoses go from the radiator to the heater and back. These hoses are subject to deterioration and should be inspected every so often.

A hose does not have to be obviously cracked, frayed, or swelled to be defective. It may look okay on the outside but be deteriorated within. Radiator hoses have a fabric or metal spring support inside to keep them open. Any hose that feels soft and mushy when squeezed is suspect.

Ask at an auto parts house for the upper and lower radiator hoses to fit your make of car and engine. Some are custom fit while others require cutting to length.

Heater hoses are always cut to length but differ in interior diameter. From the outside it is difficult to tell the difference between a ⅜-inch hose and a ½-inch hose, so you will have to depend on the parts house having a chart. If they don't, take the hose off and measure it.

Before changing hoses you will have to drain the cooling system. A hose which attaches to the top of the radiator may require only partial drainage while a bottom hose will require that you drain the radiator completely.

Diagram of cooling system showing hoses. Arrows indicate lines to and from automatic transmission. On most cars with automatic transmissions, fluid is circulated to lower radiator tank to be cooled.

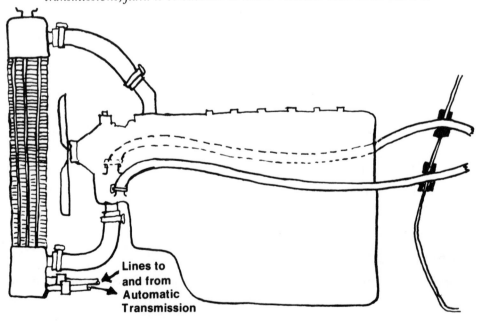

Lines to and from Automatic Transmission

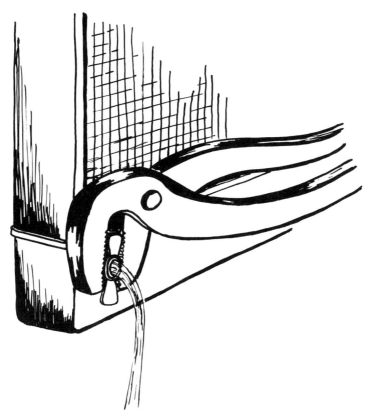

It is necessary to open petcock in order to drain radiator.

The radiator drain is a little petcock at the bottom. You can turn it with a pair of pliers. Leave the radiator cap off for faster drainage. Use a large pan to catch the anti-freeze if you want to preserve it and use it again. Don't forget to also drain the engine block when replacing bottom hoses. The block is drained by a small petcock on one side about halfway back (6-cylinder engines) or a petcock on either side (V-8 engines).

Clamps which hold the hoses on at each end are of three types. Some are simply bent pieces of heavy wire. Others, circular bands of metal which tighten with a screw. The best type are bands which tighten with a "worm-drive" sort of action. Each turn of the screw directly tightens the band.

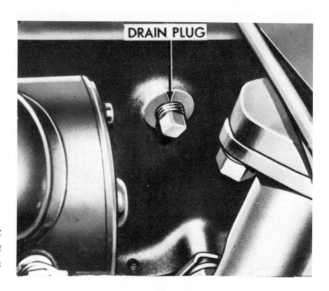

You'll find engine block drain, which is usually a small plug or petcock, on side of block. (Ford)

Wire-type hose clamps can be removed with regular pliers by compressing the ends. The job is made easier by hose-clamp pliers which have depressions in their jaws to hold the wire ends, but these are not really necessary. Other types of hose clamps simply unscrew.

When hose clamps are loose the old hose can be pulled off or levered off with a screwdriver. Re-use the old clamps with your new hose only if they are the metal strip type and not badly corroded.

The ends of the new hose may not shove on easily all the way, even after a little soap is applied. In this case go into the kitchen and boil up a kettle of water. The hose ends will fit on much more easily after they are immersed in boiling water for twenty seconds or so.

The hose clamps must be loosely fitted around the new hose before it is attached. The clamps should be fitted at least one eighth of an inch from the end of the hose and tightened so that the hose will not come off with a fairly hard tug. Do not tighten them so much that they eat into the hose unduly.

Refill the radiator and start up the engine to check for leaks.

It's a good idea to replace both radiator or both heater hoses when one goes bad. The other is likely to follow. On some cars there is also a bypass hose between the engine and water pump which may need replacing.

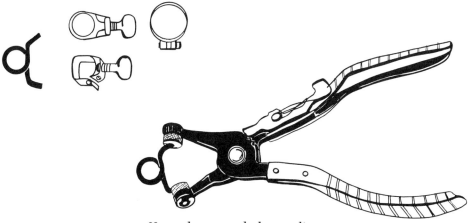

Hose clamps and clamp pliers.

THERMOSTATS

The thermostat is a small, round metal device which sits at one end or the other of the upper radiator hose, either recessed into the engine cover or the radiator itself. Look for the bolted-down housing and you should have no trouble locating the thermostat.

The thermostat may be tested by removing it from its housing and immersing it in heated water. If the plate does not open at about 195° F. (older cars may have thermostats opening at 165° to 180°—check the markings on the thermostat) then the thermostat must be replaced.

There is an easier though less precise way to check thermostats. If you suspect a thermostat is stuck closed, get the engine up to normal operating temperature and squeeze the radiator hoses to feel if coolant is circulating. A thermostat permanently open will allow coolant to circulate while the engine is still cold.

When replacing a thermostat be sure you get the correct model at the auto parts store and that it comes with a gasket which will fit. Also get a tube of gasket sealer such as Permatex Form-A-Gasket (non-hardening type).

The installation is easy. Just unbolt the housing and remove the old thermostat, noting which end is top and which is bottom. The new thermostat will go on in the same way.

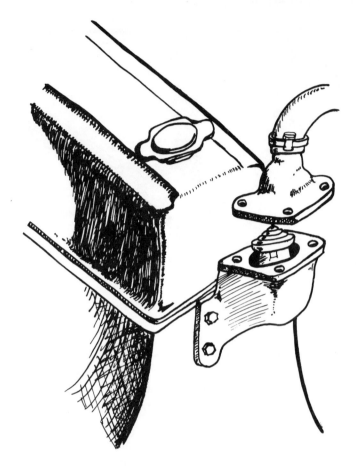

Thermostat housing on block.

A selection of thermostats. (Dodge)

Scrape the housing to remove old gasket material and wipe it clean with a rag. Then coat the new gasket with sealer on both sides and install the thermostat.

REPLACING FREEZE PLUGS

Another source of radiator coolant leakage is the freeze plugs or Welch plugs supposedly put in engine blocks to prevent cracking due to frozen coolant. They are sometimes called "expansion plugs" since they are designed to come out under the pressure of forming ice.

Actually freeze plugs are frequently necessitated in the casting of engine blocks. I suspect that they are more a production requirement than a safety factor.

When a leak has been traced to a loose freeze plug it may be possible to contain it with radiator stop-leak. Otherwise the freeze plug will have to be replaced with a new one.

The freeze plug itself is very inexpensive (less than a dollar at an auto parts house) and replacement is fairly easy if the plug is in an accessible location.

A punch or screwdriver is driven into the old plug in order to lever it out. By tapping lightly with a blunt-nosed punch the new plug is seated in the opening, then banged in with a large hammer. Be careful to keep the plug straight when you hammer on it.

The process sounds easier than it usually works out to be since most plugs are hard to get at. In extreme cases the engine must be removed from the car to get at a freeze plug. Let's hope it doesn't happen to you.

FLUSHING THE COOLING SYSTEM

Cooling systems should be flushed whenever they exhibit rust. Even if no rust is present the cooling system should be drained, flushed, and refilled with a fresh anti-freeze/water mixture every third year. Those who live in cold climates will want to do the job in fall, before the first freezing weather.

Open the radiator cap and peer in to see if the coolant looks rusty. Run a finger around the edge of the opening to check for rust and contaminants.

Since you are not saving the coolant, move the car to a place where it can be drained freely. Let the engine cool down. Then find the drain petcock under the radiator and open it with a pair of pliers (if it's a drain plug, a wrench will be required).

The radiator is not the only part of the cooling system which needs to be drained. You will also have to drain the heater and the engine block itself.

Disconnect the lower heater hose at its connection to the engine block and let the coolant run out. Then open the drain plug or petcock on the side of the engine and let the coolant drain from the engine block.

When all the coolant is out, close all drains and reconnect the heater hose. Pour a can of radiator flush into the radiator (follow directions on the can) and fill with water to just below the overflow tube.

The thermostat at the end of the upper radiator hose will be obstructing free flow of the flushing compound through the system until the engine warms up. To prevent this you have two alternatives. The best one is to remove the thermostat, scrape all traces of gasket material from the housing, and replace the thermostat with a new gasket and fresh gasket sealer when the job is done. If you are flushing the radiator as part of a regular three-year maintenance program, it's a good idea to put in a brand-new thermostat also.

The second alternative is to put a blanket over the front of the radiator so that the engine will heat up sooner and the thermostat will open faster. Be sure the blanket will not interfere with the fan when you turn the engine on.

Now put the radiator cap back on and run the engine for about thirty minutes at idle. Then drain thoroughly as you did before.

Some radiator flush compounds must be neutralized by use of another can of fluid. After draining the flushing compound connect the heater hose again and close all drains. Then dump the neutralizing agent in the radiator, fill with water, and run the engine at idle for ten minutes before draining.

Even if no neutralizing agent is necessary, you should go through this intermediate step with plain water before draining the system again and refilling with your regular anti-freeze/water mix.

Complete the job by cleaning off leaves and other debris from the front of the radiator. Also poke a wire through the overflow tube to be sure it is not clogged. As a safety precaution do not attempt to work near the radiator while the engine is running since the fan blade could easily chop a hand or finger off. Also avoid leaving tools or other objects (such as an extension light) on top of the engine or radiator. Engine vibration could cause the object to fall into the fan with disastrous results.

The operation just described is not the type of *pressure flush* a service station does. A pressure flush requires a special kind of flushing gun which uses compressed air to force flushing compound and water through the system in reverse. It can be highly effective in unstuffing a clogged radiator. Many service stations also have a radiator core cleaning gun which cleans out the air passages from the back of the core.

WATER PUMPS

The water pump keeps coolant circulating through the radiator, engine block, and heater. It is usually run by the fan belt which also powers the generator or alternator. The water pump is located directly behind the fan and the fan blades are attached to its pulley. When the fan belt breaks or starts slipping, the engine overheats.

The water pump is sealed and gets its only lubrication from additives in anti-freeze or from the rust inhibitor and water pump lubricant you put in the radiator when the anti-freeze is not changed each season.

Aside from lack of internal lubricant, the main cause of premature water pump failure is a too-tight fan belt which puts stress on the water pump bearings. Be sure your fan belt will deflect about one-half inch at its midpoint under finger pressure to keep the water pump healthy.

If you really want the water pump to last a long time, use distilled water in the radiator. The minerals and other matter in water which has not been distilled cause deposits to build up in the pump. This leads to internal erosion.

Signs of a faulty water pump are leakage and noisy operation from a worn bearing. It is rare that a water pump completely fails to operate.

Sometimes leakage between the pump and cylinder block can be caused by loose bolts or a defective gasket.

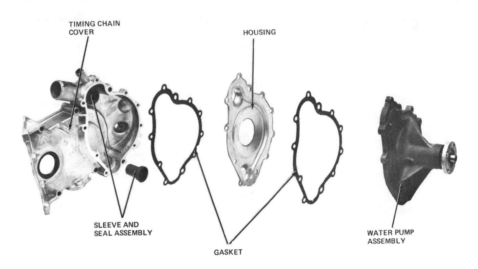

TIMING CHAIN
COVER

HOUSING

SLEEVE AND
SEAL ASSEMBLY

GASKET

WATER PUMP
ASSEMBLY

Water pump and timing chain cover. (Pontiac)

To replace a faulty water pump you can buy a new one or a rebuilt model. The cost for a rebuilt model is one-half to two-thirds the new price and it should last for the remaining life of an older car.

To replace a water pump you should start by draining the cooling system. Then take off the fan belt and unfasten the fan blades from the water pump pulley. Remove the hose connections to the water pump.

On some cars you may have to remove other parts to get at the pump. There may be a fan shroud, power steering pump, or wires and hoses. If access is difficult, you may want to consult a service manual giving the step-by-step procedure for your car.

Once you get at the water pump the job is fairly simple. Unbolt the old pump, scrape away gasket material from the mating surface of the block, coat the new gasket with sealant on both sides, and bolt on the new pump. Tighten bolts in a criss-cross pattern to avoid warping.

AIR-COOLED ENGINES

The familiar Volkswagen, the defunct but not forgotten Corvair, and the exquisite Porsche are the more familiar cars employing air to cool their engines.

There is no radiator, no anti-freeze, no circulating system. Instead there is a fan to blow air, and shrouding to ensure that it reaches the right places.

The fan is at the top of the engine driven by a belt, except that the bigger VW's (such as the Squareback) have a fan lower down powered directly off the crankshaft.

The belt is the most frequent source of trouble. It must be checked for wear and adjustment periodically. On VW Beetles and Porsches the belt tension can be given the familiar finger test. It should deflect about a half-inch at mid-point under finger pressure.

The Corvair fan belt makes two 90-degree bends and adjustment is more critical. It requires a special belt-tensioning gauge.

To tighten a VW or Porsche fan belt you undo the rear half of the generator pulley, after first jamming a screwdriver blade into the hole in the pulley side to hold it stationary, and remove washers from between the halves. The more washers you remove, the tighter the belt becomes. When all are gone you need a new belt. The washers are stored on the outer end of the generator shaft and held on by a retaining nut until you are ready to use them again.

Both VW and Corvair have a thermostat which keeps flaps closed while the engine heats up, and then opens the flaps to prevent overheating. You can see if the system is working on Corvairs by looking to see if the flaps

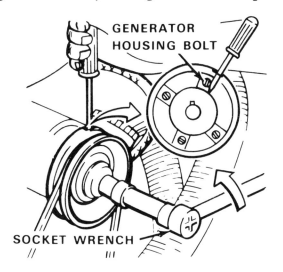

Undoing the generator pulley on VW. Screwdriver is fitted in slot to hold pulley while retaining nut is loosened with wrench. (Volkswagen)

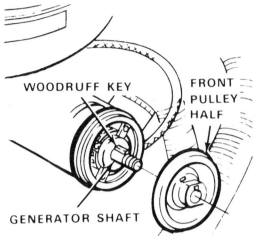

WOODRUFF KEY

FRONT PULLEY HALF

GENERATOR SHAFT

If front pulley half slides off Woodruff Key (half-moon-shaped retainer), turn shaft until groove is up. Position key in groove so top edge is parallel to shaft axis. Then seat pulley. Only rear half of pulley needs to be removed to change belt tension. (Volkswagen)

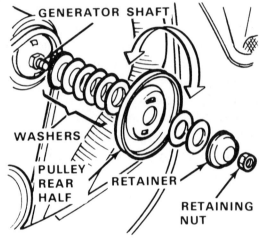

GENERATOR SHAFT

WASHERS

PULLEY REAR HALF

RETAINER

RETAINING NUT

Washers are moved on shaft to adjust belt tension. (Volkswagen)

close off the grille opening under the rear bumper as the engine heats. If not, you must take the shroud off and install a new thermostat. The thermostat is located under the bottom engine shroud.

VW flaps are under the engine and not easy to check visually. Instead, reach all the way in front of the engine shroud and feel around for a link which connects the two flaps. When the engine is cold the link should be to the left, meaning the flaps are closed. They open as the engine heats and the link shifts to the right.

The VW thermostat is under the right side of the engine. The thermostat has slots for adjustment on a bracket and should be set to touch the upper part of the bracket with the flaps fully open.

It should be noted that almost all air-cooled cars are really partially oil-cooled. They have small oil radiators which are tied in with engine oil circulation and help keep the engine cool. When an air-cooled car persistently overheats due to heavy use or engine modification, these oil coolers may be replaced with more efficient ones. The modification is particularly popular with VW Beetle owners since Volkswagens have an oil cooler which blocks off air from one of the cylinders and makes it run hotter than the others.

HEATERS AND DEFROSTERS

The design of the heater is somewhat like that of the radiator except that instead of cooling the engine it heats you.

When you push the operating lever to the "hot" position a valve opens and allows water to flow into the heater through hoses from the radiator.

There is also ducting to bring fresh air in, and a blower to circulate the heated air in the car.

When the heater fails to go on or turn-off, the cause is generally that the cable to the water valve is broken or has slipped off its attachment.

Feel around behind the heater lever under the dash and trace the cable. At the end of the cable there should be a small loop which is supposed to fit over a little lever on the interior of the firewall. This lever controls the water valve. If the cable has slipped off, find the lever and replace the loop on it.

A broken cable can be held together temporarily with a pair of vise-grips clamped down on the broken ends. Get a new cable from an auto parts store.

Sometimes a heater is not putting out enough heat because too much air is in the system and air bubbles are blocking the flow in the heater hose. Try "bleeding" the air by temporarily disconnecting the hose which returns coolant to the radiator and allowing air and a little fluid to drain out with the engine running. Then clamp the hose back on. The return hose is the one which feels colder after the engine has been running for a while with the heater on full blast. Disconnect it from the back of the radiator when bleeding.

The blower is electrically operated so troubleshooting will involve inspecting the fuse, wiring, and switch. The fuse should be marked in the fuse box. To find the wiring you can locate the blower by its sound and try to trace the wire back. A wiring diagram helps.

Another heater problem is caused by holes in the air duct. The duct (or ducts) is often made of thin rubber and tears or punctures easily. It can be mended with wide strips of duct tape and Permatex.

The defroster is just a duct (or ducts) to direct the hot air to the windshield. The duct is opened and closed mechanically by a cable. Trace it from the dash lever to see if it's operating. Look to see if the duct is plugged up.

The most common cause of defrosting trouble may very well be all the junk people put on top of the dash, directly over the defroster openings. This isn't your problem, is it?

It should be mentioned that some cars have a quite complex "climate control" system which ties in heating and air-conditioning action to keep the car at a pre-set temperature. You need a service manual to effectively troubleshoot this type of system.

AIR CONDITIONING

The air-conditioning system is based on circulation of a gas called Freon 12. When this gas is compressed enough it turns to a liquid. Then, when the pressure is lowered, the liquid Freon boils like crazy and becomes a gas again. In the process it absorbs a great deal of heat from the air in the car.

The system has a belt-driven compressor which raises the pressure of gas by compressing it. The gas then goes to a condenser which turns it into a liquid. The liquid Freon flows to a receiver/dryer where any water condensed with the gas is absorbed by silica gel. The next stop is an expansion valve which meters the Freon passing into an evaporator. In the evaporator the Freon is free from high pressure and begins to boil, becoming a gas again. This is the step where heat is absorbed from the car. Finally the hot gas from the evaporator goes back to the compressor and the cycle begins all over again.

There is not very much you can do to service the system without special tools and a manual. The Freon 12 itself is a dangerous gas and safety glasses must be worn when working with it.

There are a few things you can do to keep the system in good shape. The compressor drive belt should be checked for tightness (about one-half inch deflection at mid-point under finger pressure) and belt condition. The condenser should be kept free of dirt by scrubbing the front part with a brush and using a vacuum on it when you go to the car wash.

Refrigerant level may be checked by running the air-conditioner for fifteen minutes at the coldest position and then looking at the sight glass which is usually on top of the receiver/dryer unit. (Some are on the condenser itself.) Bubbles in the sight glass mean that the refrigerant level is down and the system needs to be recharged. Take it to a garage which has the equipment.

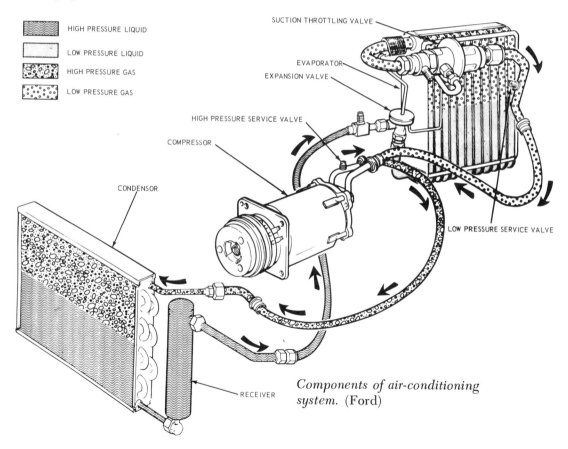

Components of air-conditioning system. (Ford)

To check on whether your air conditioner is working efficiently, use a thermometer to determine the outside air temperature and then to read the temperature inside the car after the air-conditioner has been operating in the coldest position for fifteen minutes. The air in the car should be at least 15° cooler. Air-conditioners have to do less work in dry air than in moist air so their efficiency will vary somewhat with the humidity. However, a 15° differential is the minimum you should expect.

Factory-installed air-conditioning is usually tied in with the heating system and there are often little doors opened by vacuum pressure triggered by a thermostat. If you can locate them on your system, vacuum lines can easily be checked for output and faulty thermostats replaced.

Add-on air-conditioning systems are not necessarily designed with your car in mind and installing one may lead to problems. The most usual complaints are over-heating of the engine and putting too heavy a draw on the charging system. Check an add-on as soon as you have it installed by going on a long drive; preferably in hot weather and over mountainous terrain. If the engine overheats, try a higher-pressure radiator cap (assuming one was not put on during the installation). Installing a plastic expansion tank to give a closed cooling system and getting a fan with more blades may help. Otherwise you will probably need a larger radiator or an auxiliary oil radiator.

Test the electrical load of the air-conditioner by running it while you turn your lights, radio, and electric windshield wipers on. If the ammeter reads "discharge" with the engine at idle (on cars with alternators—the same test should be made while driving at 15 m.p.h. in a generator-equipped car) you are either going to have to restrict yourself to operating fewer electrical accessories simultaneously or get an alternator with a greater output.

3

Gas and Exhaust Systems

● As you do more work on your own car, you will soon find yourself looking around for parts and comparing prices. With a little ingenuity, most parts can be obtained for a lot less than the full retail price charged by a garage.

Common items, such as fan belts, tune-up kits, batteries, headlights, and thermostats, can be found at department store automotive sections and at accessory stores such as Pep Boys and Western Auto. Their prices are substantially below what you would pay at the parts department of a car dealer.

Professional parts houses—in the business of supplying mechanics—carry a large stock of more specialized items. If you want a new exhaust pipe for your Studebaker Champion, or even a starter for a Ford Pinto, you will have to deal with a big parts house. Buy through a mechanic, if you can, to get the professional discount.

When you are not in a hurry, most parts can be ordered by mail from catalogs put out by five or six giants in the automotive mail-order business. (See Appendix D for addresses.) Foreign-car owners may find that some parts are more easily available by mail than through local suppliers.

Many parts are interchangeable among various makes and models of cars. A number of parts are shared by both luxury and economy cars and prices are cheaper when ordered for the lower-priced model.

Foreign cars from various countries often share components from a common supplier. For instance, Lucas (Great Britain), Ducellier (France), Bosch (Germany), and Marelli (Italy) supply the electrics for most cars made in their respective countries. A Lucas generator used on one British car may be the same model as that on another.

Glenn's Foreign Car Repair Manual or Chilton's new two-volume set with the same title are good sources for the model numbers of major components on most foreign cars. They also list small parts—spark plugs and light bulbs, for example—which are duplicated by American manufacturers under a different designation. The same information is in many service manuals.

The best source for listings of interchangeable parts on American cars is the Hollander manual kept by many large wrecking yards.

A wrecker or junkyard owner may be in the scrap-metal business, but most of the larger yards make their main profit from salvaging car parts. They test, catalog, and store used components and many will even give a guarantee on parts they sell. Quite a few wreckers have a shortwave hookup with other places in the area and will call around for parts they don't have in stock.

Parts houses frequently carry rebuilt components which are priced in between a new part and a used one from a wrecker. Rebuilts are components that have had worn parts replaced with new ones. There is no standard for how thorough a rebuilding job is done, so you must depend upon the reputation of the supplier. Rebuilts sold by professional parts houses are generally quite good; they are often guaranteed.

Whether your best buy is a new part, a rebuilt, or a used part from a wrecker depends primarily on the age and condition of your car.

Some parts are not worthwhile buying used. Carburetors, fuel pumps, brake parts (with the possible exception of drums), clutches, and other components which wear comparatively rapidly are a poor risk. Rebuilt spark plugs are an abomination. Engines and transmissions can be a gamble, but the price may be attractive enough to warrant it.

Rebuilts are almost always worth buying for an older car so long as they are significantly less expensive than new parts.

Instead of buying a rebuilt part, you can buy a rebuilding kit and do the job of disassembling and replacing parts yourself. Complete instructions usually come with the kit.

If your carburetor goes bad, for example, you can buy either minor or major rebuild kits. The minor kit (sometimes called a zip kit) will contain only the quickest-wearing parts such as a throttle valve, needle and seat, accelerator pump, and gaskets. The major kit will contain practically anything which could possibly be worn.

Rebuilding components is a good way to learn how things are put together, but the savings will rarely repay your time and effort. A complete rebuilt will cost little more than a major parts kit.

CARBURETORS

The carburetor receives fuel from the fuel pump, mixes it with a regulated amount of air, and sends a controlled flow to the cylinders via the intake manifold.

As fuel is pumped to the carb, it is held in a float bowl. When the fuel level rises, a float rises with it. A rod connected to the float controls the opening and shutting of a needle valve. As the level rises, the needle valve shuts and no more fuel flows in. Then the float goes down, the needle valve opens, and more fuel comes into the chamber. The principle is just like the float in a toilet tank.

The fuel is sucked into the throat of the carb by the venturi principle. In the carb throat is a venturi, or tube, which is narrowed at the middle. As air passes through this tube it is speeded up at the area of constriction and

Fuel moves from vent in float bowl (arrow) to air horn in carb by pressure induced by venturi constriction. Arrows in air horn indicate incoming air flow.

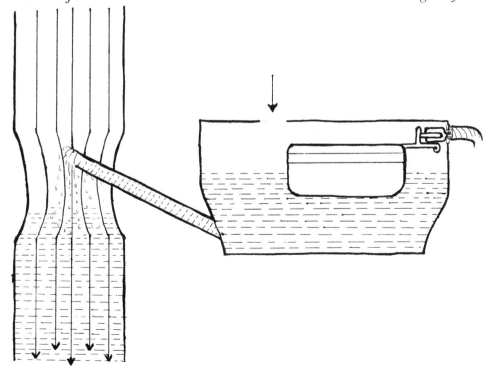

thus creates a partial vacuum. (Physics students will recognize this as an application of Bernoulli's principle.) Since the pressure in the venturi is lower than the atmospheric pressure on the gas in the float chamber, the gasoline is "sucked" into the venturi, where it mixes with air. Another partial vacuum, created by the engine's pistons moving downward on their intake strokes, causes this mixture to be "sucked" (actually pushed by atmospheric pressure) through the intake manifold into the cylinders.

The amount of mixture fed to the cylinders is regulated by the driver's foot pushing down on the gas pedal and opening the throttle plate in the carb.

This system works fine as long as the driver's foot is on the throttle. At idle there is a separate system to hold open the throttle plate slightly and keep enough fuel flowing to allow the engine to run.

Another part of the carburetor is designed to boost the flow of gas when you push the gas pedal to the floor suddenly and open the throttle plate all the way. It takes a fraction of a second for the gas flow to catch up with the air flow and the engine stumbles or hesitates. To eliminate such flat spots the carb has an accelerating pump with a separate rod going to the linkage from the gas pedal. When you floor the gas pedal, the rod forces a piston down and shoots a spurt of gas into the venturi. This causes the gas flow to keep up with the air flow.

There is also a system to compensate for the fluctuations in engine vacuum. When you mash down on the gas pedal the throttle valve to the induction system opens and vacuum is lost. To compensate, there is another valve which is kept closed by vacuum pressure, but automatically opens when the pressure is lowered. This is known as the *power valve.*

There is one more part of the carburetion system which may be on the carb itself or on the linkage or firewall. It is called the *anti-stall dashpot* and is fitted to cars with automatic transmission and some with manual shifts.

The anti-stall dashpot is designed to prevent the car from stalling when you suddenly lift your foot from the gas pedal. On an automatic transmission car the transmission does not disengage. It just pulls the engine down to idle speed right away. The carburetor is one step behind. It is still delivering a big load of gas mixture to the cylinders. This is too much for the cylinders at idle speed, so the engine stalls. The anti-stall dashpot works on the throttle linkage to keep it from closing too quickly. It works like the door closers that push a piston against air to keep a door from slamming shut.

Another part of the carb, which we will cover shortly, is the choke. It is a plate at the top of the air horn designed to shut off the air flow and richen the mixture when the engine is cold. The choke controls a fast idle cam—really just a plate with a series of detents—to increase the idle speed while the engine is being choked and thus utilize the richer mixture efficiently.

Take a look at the carb on your car and you may notice a number of hoses running to it. The gas line from the fuel pump is either a copper tube

Choke Plate

Carburetor idle system.

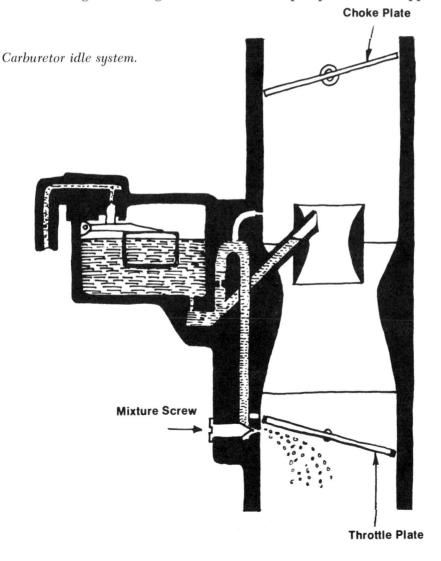

Mixture Screw

Throttle Plate

Carburetor accelerating circuit.

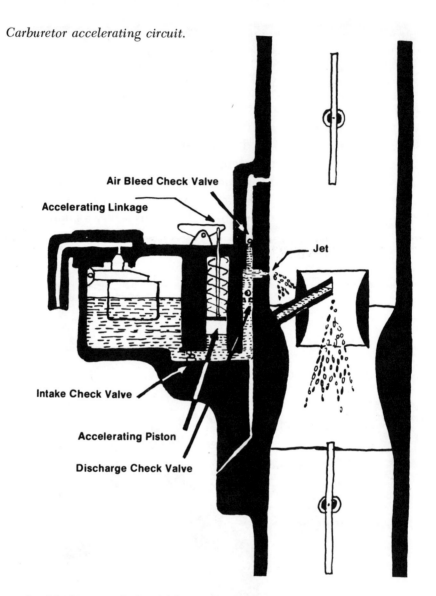

Air Bleed Check Valve

Accelerating Linkage

Jet

Intake Check Valve

Accelerating Piston

Discharge Check Valve

or flexible hose and should be easy to spot. Then you may find a hose running from the carb to a mushroom-shaped device on the side of the distributor. This is a vacuum advance mechanism which utilizes carb vacuum to advance the distributor timing at high speeds. More about it in Chapter 6.

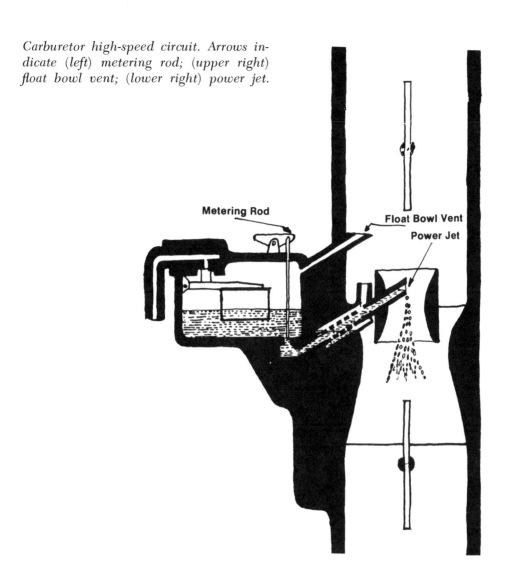

Carburetor high-speed circuit. Arrows indicate (left) metering rod; (upper right) float bowl vent; (lower right) power jet.

Metering Rod

Float Bowl Vent

Power Jet

A hose from the vicinity of the exhaust manifold to the base of the carb is a clever system for using exhaust heat to warm up the intake passages of the carb and improve fuel vaporization.

Finally, on cars of the emission control era, there could very well be an

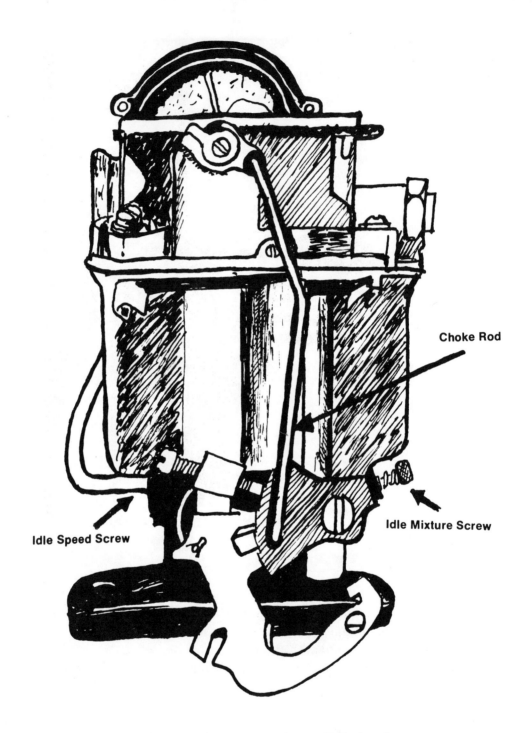

Choke Rod

Idle Speed Screw

Idle Mixture Screw

Adjustable components of single-barrel carb.

electric wire running to a little solenoid (magnetically controlled switch) on the carb. When the ignition switch is turned off, the solenoid closes the throttle plate completely and prevents overrun or dieseling. On many late-model GM cars there is a more complex version of this solenoid known as a CEC (combustion emission control) valve which works on both electricity and vacuum.

By now you are probably beginning to suspect that the modern carburetor is really a gee whiz device out of a Tom Swift book—and you could be right. Fortunately most carbs work rather reliably until their internals begin wearing out and the common solution at that point is to bolt on a rebuilt model.

There are two adjustments possible on most carbs: idle speed and idle mixture. The idle speed screw controls the amount the throttle is held open at idle. By screwing it in, you increase the idle r.p.m.

Before adjusting the idle, bring the engine up to normal operating temperature. Make sure the fast idle cam is at its lowest detent so that it is not affecting idle speed. Take off the electric wire to the solenoid, if there is one, and remove the vacuum advance line at the distributor, plugging it up with something convenient, like a golf tee.

To do the job scientifically, consult a service manual for the recommended idle r.p.m. and use a tachometer to set it. The manual will also tell you whether to set the idle with the air cleaner on or off, and will cover other conditions (tranmission in neutral or drive, for instance) which are recommended. One wire from the tach is hooked to the small coil terminal where the thin wire to the distributor is attached, while the other wire is clipped to a ground.

If you must adjust the idle and no tach is at hand, use a simple rule-of-thumb. Back the idle screw out until the engine is stumbling and begins to stall. Then screw it in again one-half turn. The engine should run smoothly, without roaring, and an auto-transmission-equipped car should idle in neutral without creeping ahead.

The idle mixture screw controls the opening in the idle passage(s). Single-barrel carbs have one mixture screw, dual-barrels have two, and four-barrel carbs still have two mixture screws since barrels three and four are secondaries which only come into operation at higher engine speeds.

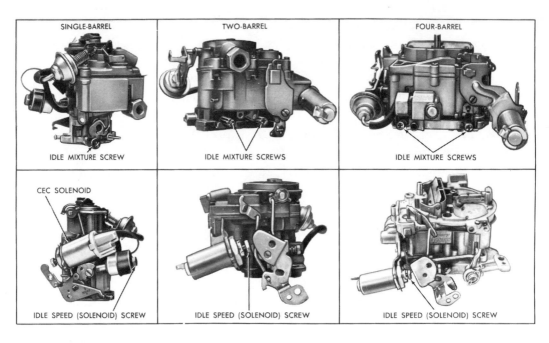

SINGLE-BARREL	TWO-BARREL	FOUR-BARREL
IDLE MIXTURE SCREW	IDLE MIXTURE SCREWS	IDLE MIXTURE SCREWS
CEC SOLENOID		
IDLE SPEED (SOLENOID) SCREW	IDLE SPEED (SOLENOID) SCREW	IDLE SPEED (SOLENOID) SCREW

Mixture screws and idle speed solenoids on some typical carburetors. (Chevrolet)

Late-model cars have a metal tab or limiter cap on the mixture screw designed to prevent tinkering. The mixture is set for minimum emissions and adjustments are supposed to be made only by those who have access to an exhaust gas analyzer.

On older cars there is again a rule-of-thumb method of setting the mixture. Screw the adjuster in finger-tight (overtightening can lead to problems) and then back it out two turns. Assuming that the idle has already been set to specifications, this should bring you somewhere in the correct range of mixture adjustment. Backing the adjuster farther out will richen the mixture, while screwing it in will make the mixture leaner and cause the engine to stall. A little fiddling back and forth is called for, especially if you have two mixture screws to play with.

A more scientific method than the above calls for the use of a patented device known as the Color-Tune, which is a poor man's version of an exhaust gas analyzer. If you have ever used a Bunsen burner in a laboratory, or welded with an oxy-acetylene torch, you will have noticed that the heat

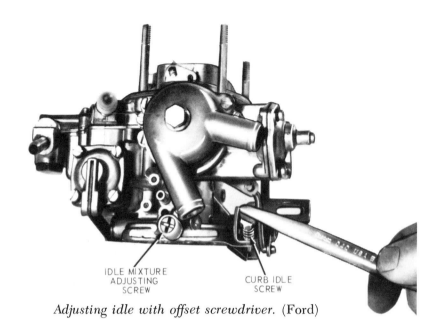

IDLE MIXTURE
ADJUSTING
SCREW

CURB IDLE
SCREW

Adjusting idle with offset screwdriver. (Ford)

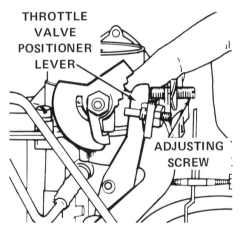

THROTTLE
VALVE
POSITIONER
LEVER

ADJUSTING
SCREW

Choke-controlled fast-idle cam, also known as throttle valve positioner. (Volkswagen)

of a flame is determined by the amount of oxygen available and that color varies from a sickly yellow through gradations of orange and blue, to an intense blue-white on the hottest flames. The Color-Tune fits a spark plug hole and has a mirror at the top which allows you to actually see the combustion in the chamber and determine whether the correct color is achieved.

While we're on the subject of gadgets, there's another one which is needed to tune cars that have two or more carbs. Various versions of this device, known as the Uni-Syn or Auto-Syn, utilize the same principle. They attach to the carb throat and measure vacuum (negative pressure, that is). To synchronize two or more carbs, the vacuum of each must be equalized.

Whether or not you decide to load up on equipment and become the neighborhood carb specialist, there is one maintenance procedure you should definitely learn to do. The chief enemy of the carburetor is dirt, so you should clean yours regularly.

The exterior of the carburetor and the linkage can be cleaned with a solvent such as Gumout and an old toothbrush to get into the tight places.

Color-Tune schematic view. Arrows show (left) connection to spark plug lead and (top) mirror.

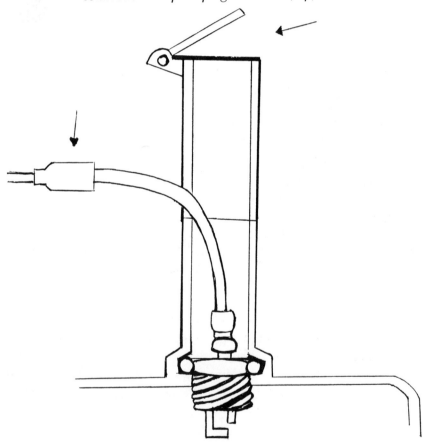

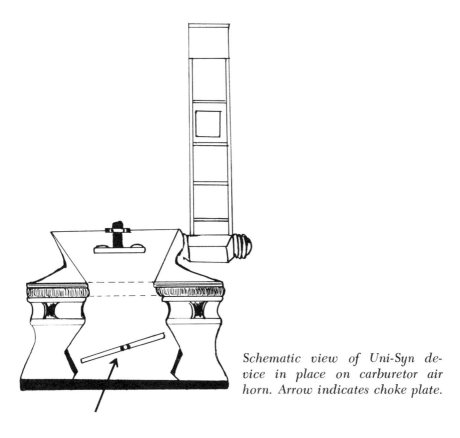

Schematic view of Uni-Syn device in place on carburetor air horn. Arrow indicates choke plate.

For internal consumption the solvent comes in an aerosol kit with a plug to put in the fuel line (after you've disconnected it) and a tube which attaches your spray can to the gas intake. Remove the air cleaner, start the engine, poke open the choke with your finger, and spritz. Place the palm of your hand over the air horn of the carb to keep the solvent circulating inside.

There is another method for achieving the same results. Buy a can of "top oil" or diesel oil and drive your car to some deserted area. Remove the air cleaner, poke open the choke, start the engine, and slowly pour the top oil in. Huge clouds of smoke will begin pouring out of your exhaust pipe and, if you didn't pick a really deserted area, fire trucks may come screaming to the scene. In any case, the build-up of deposits in your carburetor has been burned out. Re-adjust the idle, check the tightness of all the screws which hold the carb body together, and you are set for another year.

Any or all of the above procedures should be carried out only after you have tuned the ignition system following the steps given in Chapter 6. The only exception is the internal cleaning of the carb, which will undoubtedly foul your plugs a little and should therefore be done just prior to a tune-up.

Likewise, the carb should be left until last in any troubleshooting procedure. Check out the ignition system and the fuel supply to the carb thoroughly before you even think of monkeying with the carburetor's internals.

When you have definitely traced a problem to the carb, it may be time to install a ebuilt model, if that's the way you decide to go. Since the new carb goes on the same way the old one came off, you should not have much difficulty.

Be careful when loosening the soft brass pressure fitting on the gas line, since it bruises easily. Use a tight-fitting wrench, maybe even a flare-nut wrench which fits exactly, and don't be heavy-handed.

Teflon tape, the same kind a plumber uses, will wrap right around the threads on a pressure fitting and prevent leakage. But the teflon makes it all-too-easy to overtighten the connection, so go easy.

Learn to work methodically, labeling and even sketching the linkage, hoses, and wires you pull off the carb before unbolting it. Then put a new gasket on, bolt the rebuilt carb in place, and connect everything up again.

Start the engine and check for gas leakage, as well as leakage at the vacuum line, before you put your wrenches away.

Once you have become familiar with the Tom Swiftian device at the heart of your fuel supply system, the rest is comparatively simple.

Flare-nut wrench is used on soft brass fittings. (Snap-On)

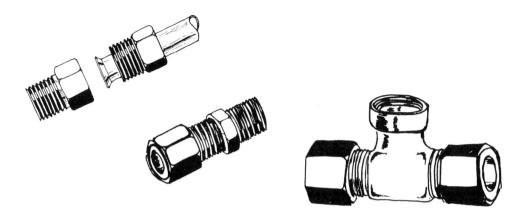

Pressure fittings are much like those the plumber uses.

CHOKES

The choke is a device for limiting the amount of air which goes to the car-buretor when the engine is cold. This causes the engine to be fed a richer mixture (more gas than usual in the gas/air mix). The richer mixture is needed because cold engines tend to condense fuel on the intake manifold and cylinder walls.

Manual chokes are operated by pulling out a lever on the dash which closes the plate at the air intake on top of the carb. As the engine warms the lever is pushed in gradually.

Manual chokes require minimal maintenance. A bit of speedometer cable lube on the choke cable once in a while will keep it from binding.

Automatic chokes are more complex. The choke plate at the top of the carb is still utilized, but the activating mechanism is a thermostatic spring which expands when heated.

Most choke coils are mounted on the exhaust manifold and expand when the manifold heats. They are connected by linkage to the choke plate on the carb.

Some choke coils are set in a well on the intake manifold, while still oth-ers are mounted on the carb itself and expand when heated fluid from the engine cooling system circulates in a chamber in the choke cover.

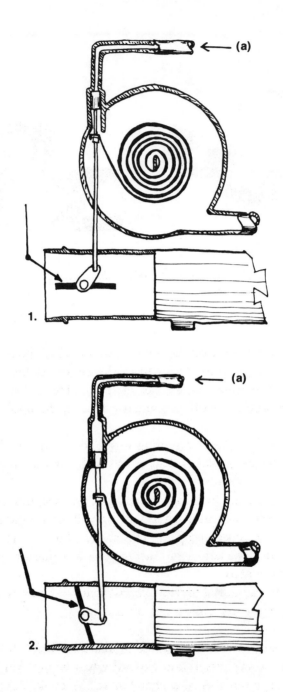

Arrows on left indicate operation of automatic choke plate: (1) intake manifold vacuum (a) tends to open plate, while (2) thermostatic spring holds it closed until engine heats up.

The choke linkage is designed so that the choke plate opens as the coil heats and expands.

After the engine starts the choke plate must be opened slightly. On manual chokes you do this yourself by pushing the choke lever in. Automatic chokes must be opened by a piston or diaphragm operated by engine vacuum. Sometimes the vacuum delivery is internal, while other designs have a vacuum hose from the engine intake manifold to the choke diaphragm.

The starting procedure for most cars is to push the gas pedal down to the floor and let it up once before starting. This closes the choke plate. If you get too much gas in the carb from repeated pumping on the gas pedal, hold it down to the floor as you try to start. This activates a choke unloader which causes the choke plate to open.

To check on whether the choke seems to be operating properly, look to see if the choke plate closes, after you depress the acclerator and release it, when the engine is cold. Then see if the plate opens slightly after you start the engine. It should continue to open as the engine gets warmer.

The only maintenance required on an automatic choke is spraying the linkage and coil with solvent at the same time you clean the carb.

To look for sources of trouble you can check the vacuum hose (if there is one) for cracks or looseness. Feel for vacuum at the end of the hose. Also note whether the engine idles more slowly as it warms up or after you jazz the gas pedal a bit on a cold engine. If not, check the fast idle cam linkage for breaks or bends.

If the choke coil mechanism seems to be corroded, look for a separate tube designed to carry heat from the exhaust manifold to activate the coil mechanism. A break in this line could be causing the problem.

Automatic chokes can be adjusted to suit the climate in your part of the country. Depending upon where the choke is mounted, adjustments are made either by resetting the choke mounting post after loosening a lock nut, or by bending the choke linkage rod. Chokes mounted on the carb adjust by loosening the cover screws and rotating the cover in the direction marked "leaner" or "richer." For the exact procedure on your car see the service manual or the section covering specific makes of carburetors in a book such as *Motor's Repair Manual.*

Cars with fuel injection do not have a conventional choking system, of course. The amount of fuel going to the cylinders is always regulated by electrical or mechanical means. Also some British cars and the Swedish Volvo (the model without fuel injection) use carburetors which operate differently and require an altered starting procedure detailed in their respective owner's manuals.

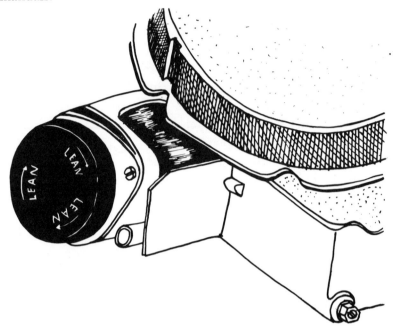

Automatic choke adjustment. Turning clockwise gives leaner mixture.

FUEL PUMPS

The fuel pump lifts or pushes the gas from the gas tank to the carburetor. Most fuel pumps are mechanical and powered by engine vacuum. A few are electric pumps and use power from the battery or generator.

By placing the gas tank above the engine, the fuel could be gravity fed and this was the method used on many early cars. On modern cars a fuel tank above the engine would be impractical and unsafe.

Mechanical fuel pumps are located on one side of the engine or another. Electric pumps are usually near the gas tank since they push the fuel rather than pull it. Some cars, such as various models of the Jaguar and the Chevy Vega, have a sealed pump buried in the gas tank.

The mechanical pump is roughly bell-shaped and bolted together all around the base where the diaphragm is located. Electric pumps have a variety of styles and can best be found by tracing the fuel line back.

When you have found that no gas is being delivered to the carburetor and traced the problem back to the fuel pump (see Chapter 7), the best thing to do with a mechanical pump is to replace it. You can repair or rebuild an old fuel pump but it hardly pays.

There are a few checks to make first, however. Obvious leaks may be caused by loose mounting bolts or a ruptured gasket. Leaks can also be caused by loose bolts around the base of the pump. Some pumps have a filter and bowl attached and this may also be loose.

The pump could be putting out too much gas, causing the carburetor to flood. This is probably due to a swollen diaphragm.

Many mechanical fuel pumps incorporate an engine vacuum booster which supplies the vacuum to run the windshield wipers. Failure of vacuum-operated wipers may necessitate replacing the fuel pump.

Electric fuel pumps are sometimes located inside tank as on Chevrolet Vega here. (Chevrolet)

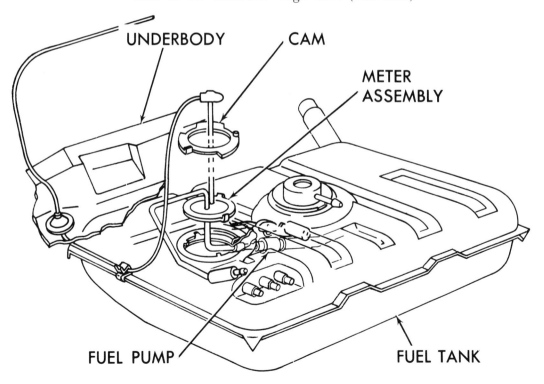

UNDERBODY   CAM

METER
ASSEMBLY

FUEL PUMP   FUEL TANK

Replacing a fuel pump is a simple job. Just detach the fuel line (it may be flexible hose held on by a hose clamp or copper line with pressure fittings), unbolt the bolts that hold the pump to the side of the engine, scrape away old gasket material, and put on the new pump and gasket, tightening the bolts and replacing the fuel line.

When you detach a copper fuel line from the fuel pump you may require two wrenches. Usually there is a fitting screwed into the pump, and the nut holding the fuel line is threaded into it. Use one wrench to hold the fitting on the fuel pump and the other to loosen the line nut.

Put a light coat of gasket sealer (non-hardening Permatex) on both sides of the new gasket before you fit it on. Run the engine and check for leaks when the installation is completed.

When an electric pump fails, the first thing to check for is a blown fuse in the fuse box or defective wiring.

You can only troubleshoot an electric fuel pump if you can get at it. Pumps buried in the fuel tank will require professional attention.

Most often the problem with an electric fuel pump is points which are burned, dirty, or stuck in an open or closed position. Sometimes you can dislodge the points by lightly tapping the pump with a screwdriver handle. The points are likely to stick again, but at least you can get home. Sometimes the points are overheated and will begin functioning again after they cool. Have patience.

If you can get at an electric fuel pump, it is just as easy to replace as a manual one. Just remove the terminal wires and hose fittings, unbolt the unit, put the new pump on, and reconnect all fittings.

It usually does not pay to repair an electric fuel pump unless you have the know-how to do it yourself. You can install new points, but something else is likely to go wrong. If the pump is more than three years old it is best to trade it in on a rebuilt unit.

See Chapter 8 for advice on the emergency troubleshooting of a fuel pump.

GASOLINE FILTERS

The main fuel filter on most newer cars is in the gas line between the fuel pump and the carburetor. It is held in place by hose clamps on either end.

In-line fuel filters are not cleaned, just replaced. This should be done at least once a year. Replacement is just a matter of loosening the clamps (with pliers or a screwdriver depending on the type of clamp) and removing the old element. The new element is placed so that its ends fit in the fuel line and the clamps are retightened.

Some fuel filters are the canister type. They may be fitted on the fuel line or on the fuel pump itself. These are unscrewed, cleaned, and the element is replaced with a new one. Some have a ceramic element which is reused after cleaning.

Most cars have two other filters for gasoline—one in the fuel tank and the other at the base of the carburetor. The former is not usually serviced. If it becomes clogged your car will stop for lack of gas and the gas tank will have to come off before the filter is cleaned. This filter stops water and its main use is to see that water does not get into the engine where it can cause harm.

The other filter, at the base of the carburetor, can be reached by undoing a brass fitting. This should be done with a wrench which fits exactly since brass fittings are soft and distort easily. This filter is a mesh screen which is simply cleaned and replaced.

In-line fuel filter (arrow). (Dodge)

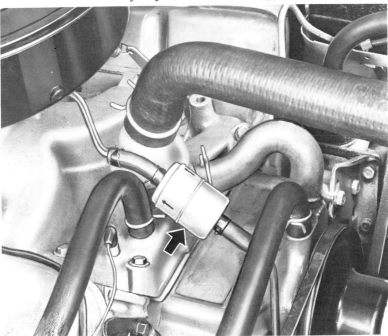

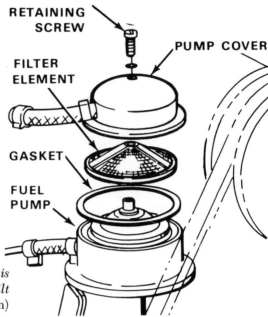

RETAINING
SCREW

PUMP COVER

FILTER
ELEMENT

GASKET

FUEL
PUMP

Where in-line gas filter is not used, filter is often built into fuel pump. (Volkswagen)

INTAKE AND EXHAUST MANIFOLDS

The intake manifold carries an air/fuel mixture from the carb to the cylinders. On V-8 engines each bank of cylinders has its own intake manifold.

Exhaust manifolds carry burnt (or unburnt) gases from each cylinder to an exhaust pipe and muffler system. Again, V-8 engines have dual manifolds and may have two exhaust pipes.

A cracked or leaky manifold, whether intake or exhaust, can be a problem to fix. Repairs with muffler sealant and fiberglass are not likely to last long. Intake manifolds can be repaired by taking them to the local welding shop and having them brazed. Even brazing does not hold too well on an exhaust manifold that's pretty shot, so you'll probably have to replace it.

New manifolds are expensive and will probably be out of stock at your local parts store. The best solution is to hunt up a manifold at a junkyard.

Getting your old manifold off and installing the new one can be a job when the bolts are rusted firmly in place. Before you start ham-handedly breaking capscrews off, you might try having the job done by a local body shop. The aggravation you save should easily be worth the price.

Remember that manifolds have gaskets also and that a leak at the gasket could be your problem.

On many exhaust manifolds there is a moveable part which is known as the *exhaust manifold heat valve* or "heat riser valve." The valve is inside the manifold and its function is to direct warm exhaust gas around the intake manifold while the engine is cold. When the engine heats up, the valve gradually opens and allows exhaust gases to pass through the pipe.

The heat riser valve helps the engine warm up more quickly and promotes smoother combustion by heating the air/gas mix in the intake manifold and thereby causing it to vaporize more readily.

The main part of the heat riser valve—the moveable plate—is inside the exhaust manifold, but the thermostatic spring, and counterweight which open and close it are on the outside. The spring is nothing but the familiar bi-metallic coil (the same type used in a thermostat) which is composed of two metals which expand at a different rate when heated and open the valve.

When the coil gets gummed up with grease and dirt or corrosion it causes the valve to stick in an open or closed position. If it sticks open, the engine heats up more slowly and may stall. When it sticks in the closed position, gas mileage goes down.

Exhaust manifold heat valve. (Dodge)

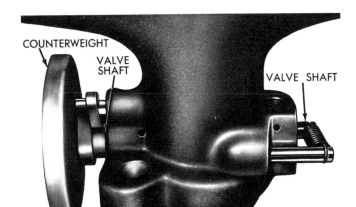

COUNTERWEIGHT

VALVE SHAFT

VALVE SHAFT

The heat riser valve is undoubtedly the most neglected part of any car. It is probable that the majority of cars on the road have a valve stuck in the open or closed position and that this is the cause of at least some of the mysterious early morning stumbling that many car owners experience. Yet the only service required is regular application of a solvent.

To locate the heat riser valve on a six-cylinder car look at the point where the exhaust manifold adjoins the intake manifold. The spring coil and counterweight may be together, or on opposite sides of the manifold connected by a shaft inside the manifold on which the plate pivots. On V-8 engines the spring coil and counterweight can be found around the junction of the manifold(s) with the exhaust pipe.

Spray the spring and counterweight with an aerosol solvent such as Gumout, which is also used to clean the thermostatic spring on automatic

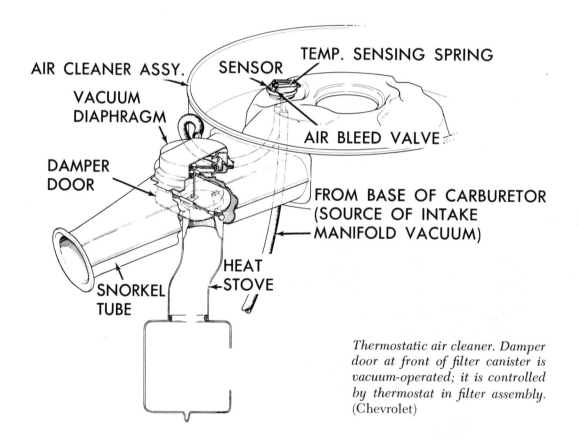

Thermostatic air cleaner. Damper door at front of filter canister is vacuum-operated; it is controlled by thermostat in filter assembly. (Chevrolet)

chokes. Manually manipulate the valve by pulling on the counterweight. If the valve is stuck, you may have to tap lightly on the counterweight with a hammer. Keep spraying and working the valve until it moves freely. That's all the service necessary. Do it whenever you change the oil on your car.

Note: On a car with dual exhausts you may have noticed smoke coming out of only one pipe on a cold morning. This just means that the heat riser valve is functioning properly and that some of the gas is being directed back to heat the intake manifold.

The only modern cars without a heat riser valve are those with a thermostatic air cleaner. This latter gizmo sounds impressive, but it is only a moveable flap funneling air into the canister which holds the air cleaner. It has the same function as a heat riser valve and is controlled by a thermostatic spring. There is a duct leading from the exhaust manifold to the bottom of the filter canister which carries heated air to the air filter and ultimately to the carburetor.

Thermostatic air cleaners have replaced heat riser valves on many late model Ford and GM cars, probably because they are not as prone to sticking. A few cars have both thermostatic air cleaners and heat riser valves.

On thermostatic air cleaners the same valve plate and linkage which blocks off the air cleaner intake moves down to block off the passage from the exhaust manifold when the engine heats up. You can check to see if the mechanism is working by watching to see whether the valve starts to open in the air cleaner intake as the engine warms. It opens gradually.

If the mechanism is not operating, spray thermostat and linkage with automatic choke cleaner and work the valve manually a little. Should this fail to make it operate properly, you will need to replace the thermostatic mechanism.

FUEL INJECTION

Fuel injection is the sophisticated modern replacement for the carburetor. Its practical application dates from aircraft engines prior to World War II, and automotive versions appeared on a few luxury production cars during the 1950s.

Like the carb, a fuel injection system mixes gas with air and sends it to

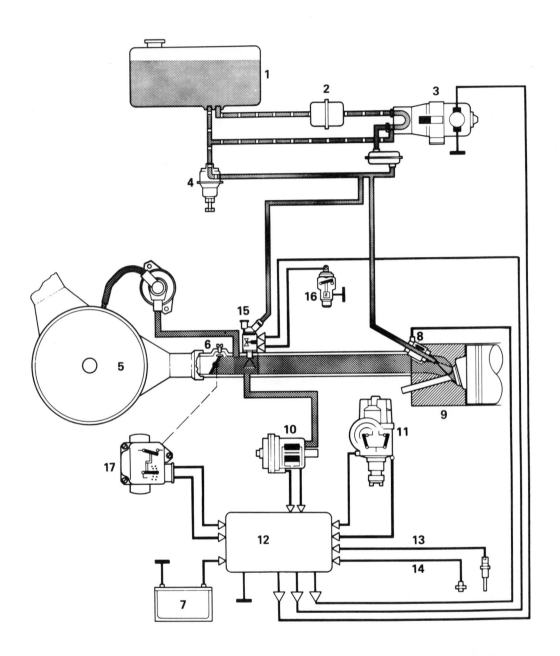

Schematic of Bosch electronic fuel injection as fitted to Volkswagen's larger cars: (1)*fuel tank*; (2)*fuel filter*; (3)*fuel pump*; (4)*pressure regulator*; (5)*air filter*; (6)*intake air distributor*; (7)*battery*; (8)*injector valves*; (9)*combustion chamber*; (10)*pressure sensor*; (11)*trigger contacts in distributor*; (12)*control box*; (13–14)*temperature sensors*; (15)*cold start injector*; (16)*cold start sensor*; (17)*throttle valve switch.* (Volkswagen)

each cylinder. However the fuel is injected into the intake manifold, or directly into each cylinder, and the amount and richness of the mixture is controlled by sensors which monitor such factors as temperature at various points, manifold vacuum, engine revolutions, and how hard your foot is tromping down on the gas pedal.

Fuel injection has become much more popular lately for two reasons. One is the introduction of electronic fuel injection by Bosch, a German firm. Electronic injection has certain advantages over the mechanical variety and is cheaper to mass produce—although still considerably more expensive than the average carburetor system. Bosch electronic injection has been fitted to the larger Volkswagens and some other German and Swedish cars. Mechanical injection is still used by a number of prestige German, French, and Italian cars.

The second reason for the popularity of fuel injection is the exhaust emission levels now required by the U.S. and some European countries. In order to come up with low emissions, smooth running at all engine speeds, and a decent power output from a relatively small engine, many European manufacturers have turned to fuel injection.

To give you an idea of how fuel injection works, let's have a look at the Bosch system fitted to the larger Volkswagens and the Porsche 914.

The brain of the system is a control box or mini-computer which can be placed anywhere on the car. On VWs it is behind the left rear quarter panel in its own little compartment. The control box is a maze of transistors, diodes, and condensers which can receive, collate, and send out information.

The injectors are at the intake manifolds and there is one for each cylinder.

Other principal parts of the system are a pressure (manifold vacuum) sensor at the left of the engine and connected to the air intake distributor with a hose, a throttle switch mounted on the air intake distributor, temperature sensors in the cylinder head and crankcase, a cold-weather starting device with a fuel jet on the intake distributor plus an electro-magnetic cut-off valve and temperature switch, an electric fuel pump, and trigger contacts in the ignition distributor.

Fuel injection control box has more than 250 transistors and other electronic components. (Volkswagen)

If you could put your brain in place of the control box, you would be getting constant messages on engine load, temperature, engine revolutions, and the accelerator position. You would have to digest this all at top speed and send out messages to the fuel injectors so that the guy who was driving the car would not notice any delay between his foot on the throttle and engine response. Since the injectors operate at constant fuel pressure, you could not tell them to shove more fuel into the cylinders at any instant; all you could do is tell them to keep shoving longer.

All this electronic gadgetry is not easy to service. The dealer's mechanics have a special Bosch electronic tester plus other miscellaneous test equipment and a complete diagnosis procedure. Perhaps this is one reason why VW is pioneering plug-in diagnosis and, to a certain extent, modular parts replacement.

Fuel injection units are generally reliable, although quite expensive to repair or replace when something goes wrong. When it comes to complexity, injection units easily outclass carburetors in the Rube Goldberg league, although they can do wonders for the performance of an emission-controlled car.

SERVICING THE PCV SYSTEM

PCV means Positive Crankcase Ventilation. It has been a requirement on new cars since the mid-sixties in order to reduce exhaust emissions (California passed the law first). To many people the PCV system is mysterious, but it is really quite simple to understand and service.

PCV valve (arrow) may be recessed in engine valve cover, as here, located near block, or in hose at air filter. (Pontiac)

The crankcase (which is where the crankshaft is located) is used to contain oil which is pumped through the car's lubrication system. The trouble is that gasoline fumes (including burned gases containing hydrocarbons) accumulate in the crankcase and condense to mix with the oil and reduce its lubricating properties. To prevent this the crankcase must be ventilated.

Old cars used a very simple means of ventilation. The oil filler cap had vent holes, allowing air to flow into the crankcase, and a lower breather tube (known as the "road-draft tube") at the back of the engine carried the fumes out. The system was quite effective but it added to the exhaust pollution by venting unburned hydrocarbons to the atmosphere. The Positive Crankcase Ventilation System, mandated by law for cars 1968 and newer, was designed to prevent this.

The PCV system may still draw air in through the oil filler cap, or the air is taken directly from the air filter. But the fumes from the crankcase are drawn into the intake manifold by engine vacuum and mixed with the air/gas mixture in the carburetor. In effect, a portion of the unburned hydrocarbons are recycled continually until they do burn.

The vacuum drawing the fumes from the crankcase is controlled at various engine speeds by a clever little gadget known as a PCV Valve. This little valve is usually screwed into the engine rocker cover (the top plate over the engine) or held into the rocker cover by a rubber grommet.

A fairly thick vacuum hose leads from the other end of the PCV valve to the spacer plate at the bottom of the carburetor. Sometimes the PCV valve is at the carburetor end of the hose rather than the rocker cover end.

If the PCV valve gets gummed up and stops functioning, the oil will be contaminated with foreign substances and will not provide adequate lubrication. There may also be rough idling, loss of power, and a high-speed engine miss since the intake manifold is not being fed the proper mixture. Pressure build-up in the crankcase could blow the engine oil seals and cause leakage.

One sign of a clogged PCV valve or stopped-up hoses is smoke coming out when you remove the oil filler cap. But there is a much more systematic way of checking out the whole PCV system.

On the type of PCV system which has a hose from the air filter to the oil filler cap, remove the hose at the air filter end (you may have to loosen a hose clamp) and put your hand over it with the engine idling to check for

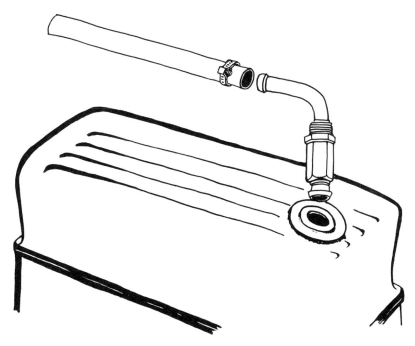

PCV valve components.

vacuum suction. Nothing? Then either the hose is clogged or the trouble is somewhere else. Pull the other end of the hose off and check for suction at the connecting tube. It may not be strong, but there should be a definite pull.

No suction at one end of the hose and suction at the other means it's clogged up. You can try blowing it out with an air hose at a gas station or buy replacement hose of the same interior diameter and cut it to fit. You may find a fine wire mesh filter at one end of the hose. Clean it in solvent and replace it. Sometimes you have to remove the wire mesh filter from inside the air filter canister before pulling the hose off.

The next step is to check the hose running from the engine rocker cover to the base of the carburetor for vacuum. (This is the first step on cars which do not have the hose between the air filter and oil filler cap.) With the engine idling, pull off the hose at the engine rocker cover end and check for suction. Then check at the carb end. Again, no suction at one end and suction at the other means a clogged hose which must be blown out or replaced.

Now check the PCV valve itself. You probably uncovered it when you removed the hose at the engine rocker cover, or else the end of the hose was attached to it. Some PCV valves pull right out from a rubber mount while others must be unscrewed with a wrench.

To test a PCV valve, shake it. A distinct rattle means the valving mechanism is functioning. Just clean the valve off and replace it. No rattle means the valve is gummed up. You can clean it or buy a new PCV valve at an auto accessory store for about two dollars. Be sure to specify the make and model of car.

The early type of PCV valve came apart for cleaning. After unscrewing it you could soak the components in kerosense, air dry, and reassemble. Later valves do not come apart. Cleaning them requires a can of special PCV solvent (the use is marked on the can) with a spout which fits the end of the valve. Squirt, and shake the valve until the interior mechanism frees up. Soaking in an ordinary solvent followed by vigorous shaking can sometimes work also.

The PCV assembly should be checked every time you change your oil and the valve should be cleaned or replaced at 12,000-mile intervals or once a year.

CRANKCASE BREATHER CAP

The part with the fancy name of crankcase breather cap is really just the cap that fits on the tube where you pour engine oil in. Sometimes it is called the oil filler cap.

On many cars this cap is filled with wire mesh and serves to ventilate the crankcase, where the oil is, and keep dirt out. The crankcase breather cap should be cleaned whenever the oil filter is changed. It is simply soaked in a solvent such as kerosene or carbon tetrachloride.

AIR INJECTION SYSTEM

To help prevent exhaust pollution many newer cars have an air injection system to force air into the engine exhaust ports. The air combines with unburnt fuel and aids in complete combustion.

The system consists of a belt-driven rotary pump near the front of the

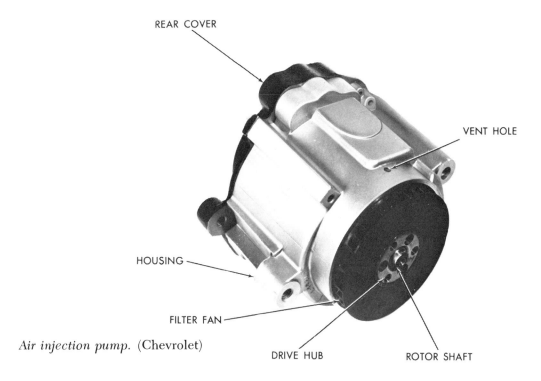

REAR COVER

VENT HOLE

HOUSING

FILTER FAN

Air injection pump. (Chevrolet)

DRIVE HUB

ROTOR SHAFT

engine which draws in air through a filter and sends it to the cylinders via an air distribution manifold. Air is collected at the manifold and distributed to the exhaust ports.

There are also two valves: a diverter valve which keeps the air from going to the ports when intake manifold vacuum rises (thus preventing backfire in the exhaust system), and a check valve in the distribution manifold to keep exhaust gases from flowing back into the injection system.

Servicing is relatively simple. First check to see that the drive belt is in good condition and correctly tensioned (one-half inch deflection at midpoint under finger pressure) and that all hoses are clamped on tightly.

Then inspect the filter. There are three types of filters used. One system uses the regular air filter on top of the carburetor and draws air from it with a hose.

Another system uses a separate filter with a pleated paper element. This is serviced just like the carburetor air filter but should need changing less often (every 24,000 miles).

A third type of filter is a plastic one behind the pump's belt-driven pulley. To change it the pulley is unbolted and the plastic filter with its fan blades is pried off. This may break the fan blades but the new filter assembly has its own blades. Place the new assembly on the pulley spindle and tighten it by bolting down the pulley.

Check out the diverter valve by pulling off the vacuum line which goes from the base of the carburetor to this valve. You should feel vacuum with the engine running. Have someone blip the gas pedal and you should feel air coming out of the diverter valve each time the gas pedal is released.

If there is no vacuum at the line, check for a clogged hose. No air when the gas pedal is released means the diverter valve is faulty or else the pump itself is not putting out any air. Investigate the latter possibility by pulling off the large hose on the pump (the one which goes to the air distribution manifold) and feeling the outlet neck with the engine running. There should be a blast of air.

The check valve, or valves (there are two on a V-8 engine) is in the hose(s) leading to the cylinder heads or on the air distribution manifold itself. Find it and push in the valve to see that the linkage which controls it operates freely. Start the engine and check for exhaust gas leakage at the valve.

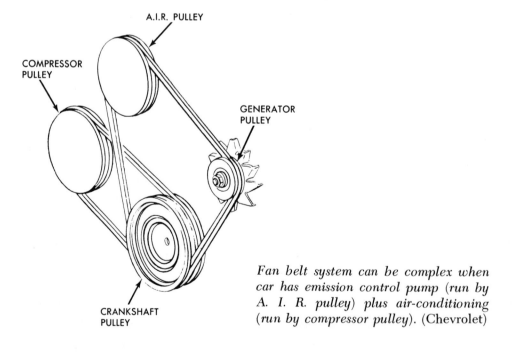

A.I.R. PULLEY

COMPRESSOR PULLEY

GENERATOR PULLEY

CRANKSHAFT PULLEY

Fan belt system can be complex when car has emission control pump (run by A. I. R. pulley) plus air-conditioning (run by compressor pulley). (Chevrolet)

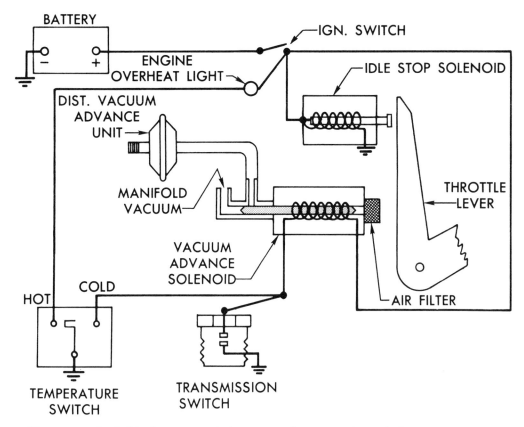

BATTERY

IGN. SWITCH

ENGINE
OVERHEAT LIGHT

IDLE STOP SOLENOID

DIST. VACUUM
ADVANCE
UNIT

THROTTLE
LEVER

MANIFOLD
VACUUM

VACUUM
ADVANCE
SOLENOID

AIR FILTER

HOT   COLD

TEMPERATURE
SWITCH

TRANSMISSION
SWITCH

The principle behind many emission control systems is to induce more complete combustion by retarding spark under various conditions. Schematic shows GM's Transmission Controlled Spark system which eliminates ignition vacuum advance when car is operating in low forward gears. System has temperature override switch. (Chevrolet)

With proper maintenance the system is pretty trouble-free. The most likely cause of trouble (aside from the hose and drive belt) is the diverter valve—an inexpensive item to replace. If the pump itself fails, you may be able to obtain a rebuilt one from an auto parts store.

REPAIRING OR REPLACING THE EXHAUST SYSTEM

Trouble in the exhaust system is usually quite apparent since the engine sound changes to a dull roar. When you have inspected the exhaust manifold, the headpipe, and the muffler for leaks, and have not found any, there's a little trick you can use to do a fast check.

Stuff the exhaust pipe loosely with a rag. Then start the car, take off the air cleaner, and pour some top-oil down the carb throat. Black smoke will come out at the point of the leak.

As long as the headpipe and muffler are not falling apart from rust, it pays to repair a leak rather than replace the whole system. The repair won't stem the incipient cancer, but it will keep you rolling for a time.

Scrape away the rust at the point of the leak, sand down to bare metal, and apply some muffler sealant such as Holt's Gun Gum. Wrap fiberglass tape (or the "muffler bandage" sold at auto accessory stores) around the leak and cover with another coat of sealant.

For extra protection you can cover the repair with a beer or soda can. Cut the bottom and top off the can and slit the side with a pair of tin snips. Wrap the can around the pipe over the patched area and use two large ratchet-type hose clamps to fasten it on tightly.

Replacing an exhaust system is a real fun job for masochists. It begins with the joy of carting a big, long headpipe back from the parts store with the end sticking two feet out of your car window, and keeping a wary eye out for cops. Then you collect your chisels, hacksaw, and rubber mallet and put the car up on jackstands.

Replacing the entire system is easier than putting on only a new muffler or pipe. In the latter case you're going to have to mate the shiny new parts with the rusty old bunged-up ones, usually requiring a little surgery.

At the muffler shops they have all sorts of gadgets like pipe mandrels, to expand the ends of pipes so that a new pipe will slide in and mate easily, chain wrenches, pipe cutters, power chisels, and even electric pipe joint heaters so that heat can be applied to loosen the rusty joints without using a torch near the gas tank.

Lacking most of these aids you can still do the job as long as you don't mind skinning a few knuckles. The items you are most likely to need are the pipe mandrel, a flat chisel that can be used to drive the muffler off the headpipe, a hacksaw to substitute for a pipe cutter, and a large pipe wrench of the plumbing variety. You also need the right mental attitude enabling you to think nice thoughts as the particles of rust float down on your head and oil drips down your shirt collar.

To do an effective job you will need support brackets and joint clamps matching those on the original system. It's a good idea to examine the installation carefully before you visit the parts store.

Penetrating oil and a chisel will help with the stubborn nuts. If the tailpipe or headpipe won't slide off the muffler (sometimes the system is welded), use your hacksaw. Slitting the pipe joint with a chisel can help too.

When you are re-using parts of the old system, clean joint areas with sandpaper. Always use a new gasket between the headpipe and the exhaust manifold. Connect the whole system and be sure that the alignment is correct before you tighten any clamps. Use exhaust system sealant at all joints and then tighten the clamps over it. Be sure all the hangers are secure.

Since exhaust system leaks can be dangerous, test for leakage after the job is complete. Use the rag stuffed in the exhaust pipe and top oil method.

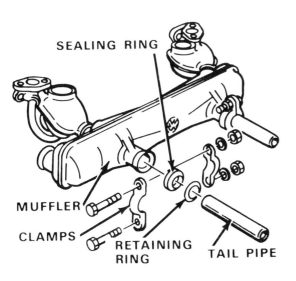

Typical exhaust system on rear-engined car. (Volkswagen)

4

Electrics

● Now that you have become more familiar with the under-hood area of your car, you have probably become quite conscious of how dirty it is. A mixture of grease and dust seems to cover everything. Whenever you touch anything your hands wind up covered with grime.

Engines do get dirty in the natural course of operating the car. But the main reason for an engine being really cruddy is that you've never bothered to clean it. Even people who make regular weekly visits to the car wash rarely think about cleaning up the engine compartment.

The reasons for cleaning an engine and the associated parts under the hood are not just sanitary. Dirt and grease on an engine act as an insulator. They tend to prevent engine heat from radiating and can raise the operating temperature of the engine significantly. Dirt can be a contributing factor—even a sole cause—of engine overheating.

Another advantage of having a clean engine is that you can quickly spot oil leaks.

Dirt is even more harmful to the electrical system than it is to the engine. Dust attracts moisture and moisture can cause a battery to discharge or the ignition system to misfire.

There are commercial steam cleaners who will thoroughly clean an engine for $6 to $8. However, you can do a more careful job yourself for only a few cents. The only equipment necessary is an aerosol can of degreaser, kerosene, rags, small plastic bags and rubber bands, and some tags to go on your spark plug wires when you disconnect them.

Begin by running the engine for a while to get it up to normal operating temperature. Most spray degreasers are designed to work best on a hot engine (check the instructions on the can). Avoid smoking as degreaser is inflammable.

Then pull off the spark plug wires where they connect to the distributor cap. Take them off one at a time and tag them so that they can be replaced correctly. Pull off the wires to the coil also.

Cover the distributor, coil, alternator, and oil filter with plastic bags held on tightly with rubber bands. Unclamp the battery cables to avoid electrical shock while working on the car.

Continue by removing the air filter canister and fastening a bag or piece of plastic tightly over the top of the carburetor to prevent water from getting in.

Spray degreaser thoroughly over the engine and associated parts. Give it twenty minutes or so to do its stuff and then wash it off with a garden hose. Don't be afraid to use high pressure where needed.

Use kerosene and rags to clean other parts which are still greasy. An old toothbrush will help clean in tight places. A household solvent such as *Mr. Clean* also makes a good cleaning agent.

Dry the exterior part of the spark plugs and the spark and coil wires. Remove plastic bags and reconnect wires. Replace the air filter.

CLEANING AND TESTING THE BATTERY

Now that you have done a thorough job on the engine, proceed to clean and test the battery and its connections. A half hour's work here will help ensure easy starting all winter long.

Tools needed are a battery hydrometer, which looks like a giant eyedropper, a small adjustable wrench, rags, a scraping knife, sandpaper, and vaseline. For corroded battery terminal connectors, which refuse to come off the terminals, a small pulling tool (battery terminal puller) may also be needed. A hydrometer costs about $2 and a terminal puller $2.50.

Most terminal connectors are on top of the battery and are of the clamp type. They loosen by undoing a bolt at the side. There are also the older spring-type terminals and new-model batteries with side terminals and connectors which screw in. The latter are the simplest to remove.

Clamp or spring-type terminal connectors which seem stuck can often be freed by gentle prying. Easy does it since forcing can break the terminal's interior contact with the battery plate connectors. Stubborn cases will require the battery terminal puller, mentioned above, which works by the action of a screw forcing two parts away from each other.

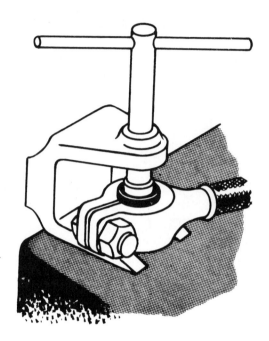

Battery terminal puller. (Owatonna)

Scrape the terminals and the insides of the terminal clamps with a knife to remove hard deposits. Use rough sandpaper to brighten the contacts. Wipe the battery with a rag to remove grease and dust.

Use another rag to clean the battery thoroughly with ammonia. It is important to prevent the slightest bit of ammonia from getting inside the battery so tape up the vent holes in the cell caps before you begin.

Once you have cleaned the terminals and contacts and removed any whitish sulfate deposits from the exterior of the battery with ammonia, dry the battery thoroughly. Replace the battery post clamps (terminal connectors). Tighten the connector bolts so that the clamps cannot be moved by hand pressure.

Cover the clamps and posts with a light coating of petroleum jelly. Then trace both battery cables to their connection at the other end, loosen and clean them, and make sure the connection is secure.

Check to see that the battery hold-down bolts are tight and that the battery is being held rigidly in place. Be careful not to use great force and overtighten the bolts since this could cause the battery case to crack.

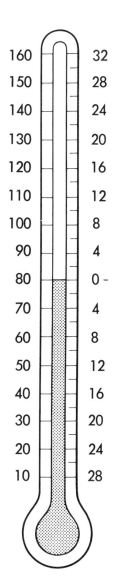

160	32
150	28
140	24
130	20
120	16
110	12
100	8
90	4
80	0 -
70	4
60	8
50	12
40	16
30	20
20	24
10	28

Correction chart for battery hydrometer. Figures on left show degrees Fahrenheit; figures on right show gravity points to add or subtract from hydrometer readings. When reading is below 80° (shaded area), subtract points shown directly opposite. When above 80°, add points. (Dodge)

Use your hydrometer to check each cell in the battery for the specific gravity of the electrolyte. The specific gravity of any fluid is the weight of that fluid in relation to the same volume of water. The electrolyte in a fully charged cell should be 1.260 to 1.280. This means that the electrolyte (acid/water mix) is 1.260 to 1.280 times heavier than the same amount of water. A specific gravity near 1.225 in all cells indicates that the battery can usually be brought up to snuff by charging. A reading of below 1.150 in any cell means that you need a new battery. In between these two readings the battery is marginal. Don't test the battery after you have just added water (the reading will be low) or after completing a high-speed run (when the reading will be inflated). The best time to test the battery is after the car has been run a bit and has had a chance to cool down.

Use the rubber bulb on the hydrometer to suck up just enough water to

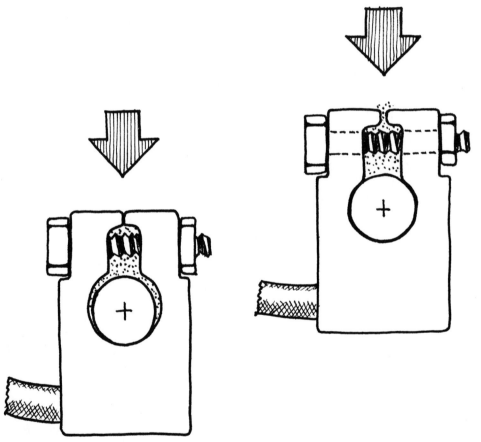

Overtightening, at left, will loosen terminal clamps. Gap must be left between battery terminal clamp fingers, as at right, for best contact.

float the indicator. Then hold the hydrometer up and read it at a level position. *Caution:* Be sure not to smoke when testing a battery or to use a cigarette lighter to view the electrolyte level in the cells. Batteries generate explosive gases.

For greatest accuracy, hydrometer readings can be temperature calibrated. They are designed to be accurate when the electrolyte is at 80° F. Add .004 for each ten degrees hotter and subtract the same figure for each ten degrees below 80°.

Those who are not purists will be content to take a reading which is not temperature corrected and even to use the type of hydrometer which uses floating balls to indicate the rough level of charge rather than giving the exact specific gravity.

When a car is used mostly around town and has quite a few electrically operated accessories, the generator may not put out enough juice to keep the battery fully charged. Alternator equipped cars have different problems. The alternator puts out enough current to supply all needs even when the engine is idling. However, alternators cannot replenish the charge used up to run accessories with the engine shut off. When driving habits result in a chronically low battery, a small, "trickle-type" battery charger is the answer. It is plugged into the electric supply at home and gives the battery a slow charge overnight through wires connected by a clip to each of the battery terminals. Chargers in the 6- to 10-amp range are most effective.

FUSES

The fuses in a car serve the same purpose as those in your fuse box at home, although they are smaller and constructed differently. Fuses are little glass tubes with a metal cap at each end and a metal element inside. Some cars, such as Volkswagens, use ceramic fuses.

When an electrical circuit becomes shorted or overloaded due to loose or faulty wiring or some other malfunction, the fuse blows.

Fuses are a safety factor. They keep the electrical accessory itself from burning out and may prevent a fire. You can replace a blown fuse easily, but when it blows again look for the true source of the trouble.

A box of assorted-size fuses costs about a dollar and should be kept in your car at all times.

The fuse box on most cars is located under the dash in the vicinity of the steering wheel. A few cars have a fuse box in the glove compartment or mounted on the firewall under the hood. Most fuse boxes have nice little labels showing the circuit each fuse is wired into. When an electrical accessory stops operating look to the fuse box first.

Headlights and windshield wipers normally use a circuit-breaker rather than a fuse. A circuit-breaker breaks an electrical connection when there is

Fuse block. (Dodge)

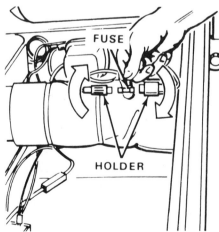

Where in-line fuse is used, it can be found by tracing wire back or consulting wiring diagram. (Volkswagen)

an overload and then automatically makes the connection again. When your headlights or windshield wipers cycle on and off, you can see the circuit-breaker in action. Circuit-breakers are usually trouble-free so the cycling action indicates a short in the wiring or an overloaded circuit.

A few electric accessories have an in-line fuse on some cars. When you find no fuse in the fuse box for that circuit, trace the wiring back to find an in-line fuse.

REPLACING BULBS

All the little light bulbs in a car, from the stop and turn signal lights to the tiny bulbs behind the instrument panel, are pretty simple to replace so long as you can get to them.

Bulbs come in 6- and 12-volt varieties, put out varying wattages, and have bases known as miniature bayonet, bayonet, double contact bayonet, or wedge type. A few bulbs are the cartridge type.

Bayonet base bulbs are loosened by being pushed toward their sockets and twisted counterclockwise.

A bulb which is difficult to free is probably corroded around the base.

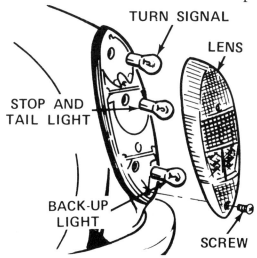

Older cars frequently had two tail light bulbs, one having a dual filament. New style (shown) has three separate bulbs. (Volkswagen)

There are special pliers made to grip bulbs but these are not really necessary. If you break the bulb carefully (to avoid cutting yourself), you can free the base with regular pliers.

The base of the bulb has a number designating the type. Request the same number bulb from the parts store. Some bulbs have been replaced by different types which fit the same socket. (For instance, the older 1034 stop and turn signal bulbs have been superseded by 1157's.) The man behind the parts counter should have a list to consult.

Bulbs in foreign cars have a different numbering system but each has its American equivalent. Consult *Glenn's Foreign Car Repair Manual* or a service manual for the American replacement number.

When the socket is corroded, clean it with a wire brush (a smaller version of the brush sold to clean battery terminal clamps). Use penetrating oil as a solvent and then wipe it off thoroughly.

Most bulbs are reached by taking off an outside lens held on by screws. When replacing the lens, avoid overtightening the screws or it may crack. Some lenses are held by a screw from the back of the lamp housing, and a few rubber coverings which must be peeled back to free the lens.

Bulbs with fixed lenses are reached from the inside. For instance, tail lights may be replaced through the trunk compartment. Sometimes it is necessary to pull the whole assembly back to a point where you can reach the socket.

REPLACING SEALED-BEAM HEADLIGHTS

Cars with four headlights have single-filament, Type 1, sealed beams inboard and Type 2, double-filament (high and low beam) lights outboard (or Type 2 above and Type 1 below). Both types are replaced in a similar manner.

As with many other parts, sealed beams are available at an auto accessory store. Type 1 and 2 headlights are universal replacements fitting most cars, with the exception of a few older models.

To replace a sealed-beam headlight, first remove the exterior chrome bezel (ring) which is usually held on by Phillips screws. Two other Phillips screws are frequently set into notches in the bezel but do not serve to hold it on. *Do not touch these two screws.* They are for headlight adjustment and you can replace a sealed beam without adjusting it.

Underneath the chrome bezel is a plain metal ring held on by slotted

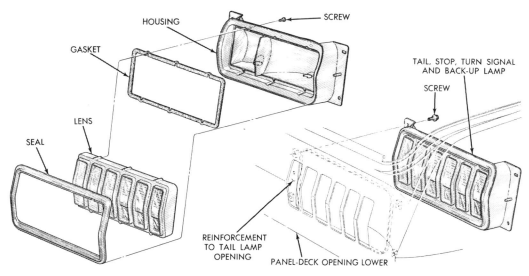

HOUSING

SCREW

GASKET

TAIL, STOP, TURN SIGNAL
AND BACK-UP LAMP

SCREW

LENS

SEAL

REINFORCEMENT
TO TAIL LAMP
OPENING

PANEL-DECK OPENING LOWER

Many tail light housings have outside retainer as shown. Some must be reached through trunk. (Dodge)

To remove sealed-beam headlight unit, exterior chrome and retaining ring must first be taken off. (Ford)

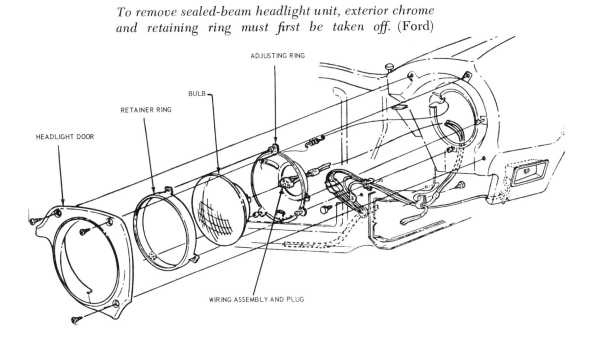

ADJUSTING RING

BULB

RETAINER RING

HEADLIGHT DOOR

WIRING ASSEMBLY AND PLUG

115 · *Electrics*

screws. Take it off and you can pull the sealed beam out enough to get at the plastic plug behind it. Some headlights have retaining springs which must be pulled with a notched tool. A good substitute tool is a bent piece of coat hanger wire.

On Type 2 lights there are three wires going into the plastic plug (high beam, low beam, and ground), while Type 1 lights have two wires (single beam and ground). Note where each wire goes and then carefully pry the plastic plug off with your screwdriver. Connect the new sealed beam, being sure that the position marked "top" (sometimes it's an arrow) is at the top. Before you replace the retaining ring be sure the headlight works on both beams (or on only one—if it's a Type 1).

ADJUSTING HEADLIGHTS

Headlights are easy to adjust but difficult to adjust right. The more trouble you take, the better the job.

Or should you adjust your own headlights at all? In a few states it's actually illegal. Presumably the law was passed to guard against headlights being adjusted to shine directly into the windshield of oncoming cars.

Professional mechanics have a headlight aimer that attaches to the lens of sealed-beam units and allows them to adjust the headlights in a jiffy. You'll have to work harder.

Two Phillips screws notched into each headlight rim control vertical and horizontal adjustment. If you don't see any adjusting screws on your headlights, take off the outer chrome ring and you'll see them underneath.

You will have to adjust the headlights so that the "hot spots" of a Type 1 beam carry straight for twenty-five feet and the low beams of a Type 2 drop about three inches down and two inches to the right of dead center at the same distance. Don't worry about the Type 2 high beams. They are automatically set right when the low beams are adjusted.

To make these adjustments you will need a spot where your car is precisely level and the beams shine on a wall or garage door twenty-five feet away. Be sure the car's tires are inflated correctly and there is no excess weight in the trunk compartment. You can check the level with a carpenter's tool used for that purpose.

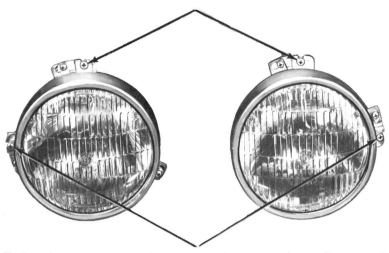

Headlight adjusting screws: those on top (see arrows) usually control vertical adjustment; those on sides (see arrows) control horizontal adjustment. (Dodge)

On the aiming wall you will have to mark a horizontal line that is level with the ground and at the same height as the centers of the headlights. A vertical line is then marked to line up with the exact center of the car's hood. This is best done by sticking a piece of clay or some other aiming object on the central point between the headlights and putting another aiming point at the mid-point of the trunk's width. (On cars with a center-mounted hood ornament, this will do as a front sight.) Have an assistant move a yardstick while you "aim," using the improvised sights. When the yardstick is in line with the sights, make your vertical line. (Pieces of string and thumbtacks are used to make the line.)

You will still need to mark two more vertical lines. These should be parallel to the center line and the exact distance from it that the center of the headlights is from the mid-point of the width of the hood. Where these two vertical lines intersect the horizontal line is where the hot-spots of Type 1 headlights should center. The low beams of Type 2 headlights should have the central point of their hot-spots three inches down and two inches to the right.

There's just one catch. On cars with four headlights the Type 1 beams are inboard and the Type 2 are on the outsides (or the Type 2 above and the Type 1 below). Since the beams are placed differently, they will require that you draw two separate sets of lines.

Remember that whenever you put a lot of weight in the trunk compartment the rear end of your car will go down and the front end will go up. This will raise the headlights and throw their aim off until you take the weight out. If you always travel with a load in the trunk, or you are making a long trip with a trailer hitched on, re-aim the headlights to compensate for it.

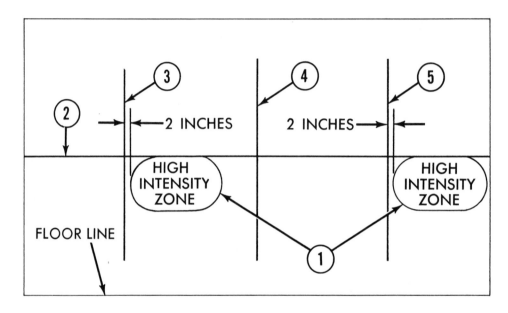

Pattern for adjusting low beams. (Dodge)

FLASHERS

The turn signals on all cars and the emergency flashers on cars after 1967 are controlled by small electromagnetic switches which cycle continually on and off. The switch mechanism, known as a flasher itself, simply plugs into the appropriate circuit.

When turn signals or emergency flashers go on but fail to cycle on and off, it usually means a tail light bulb is burned out and must be replaced.

However, when flashers begin to function very slowly and weakly, the flasher mechanism must be replaced.

Auto parts stores will carry both regular and heavy-duty flashers sized to fit your car. They look somewhat like radio tubes except that they are made of metal. Unless you have extra flashing lights (on a trailer for instance) get the standard variety since heavy-duty flashers will not function correctly.

Turn signal flasher unit.

There are separate flasher mechanisms for the turn signals and emergency flashers. Both are located under the dash on most cars. A little hunting up under the instrument panel or under the mid-dash area should turn up the flasher mechanisms.

On older Fords, including Falcons, Comets, and Fairlanes, the flashers are not readily visible since they are rather high up behind the panel. Getting at them will require small hands and patience.

Thunderbirds before 1970 and older Cougars have flashers located in the trunk compartment. New Cadillacs have turn signal flasher mechanisms on the underside of the steering column cover. Some Ford products (including models of Lincoln and Mercury) have flashers in the vicinity of the glove compartment.

WIRING

When any electrical accessory stops working it's natural to suspect that either the fuse for its circuit is blown or the wiring is defective. After checking the fuse the next step is to troubleshoot the wiring.

Wires loosen, break, or fray due to heat, vibration, and rubbing. A wire may even break inside its insulation making detection difficult. It is not easy to trace a wire to its origin since you are liable to encounter junction blocks, groups of wires twisted together in a harness, intermediate switches, and even other accessories which the wire serves. Most wires are color-coded to keep from confusing them, but color-coding is often not much help unless you have a wiring diagram showing which color wire goes where.

When wires do break or fray, they may touch an electrical conductor (in which case the current jumps to the conductor and the circuit is shorted) or simply leave an open circuit. Unlike a short, an open circuit won't blow a fuse. But it certainly will stop the accessory from operating.

To troubleshoot wiring you need a battery-powered test probe. It has a pointed end which can be stuck through the insulation to contact a wire, a bulb which is powered by its own battery, and a lead wire with a clip to make it easy to ground.

Since the test light has its own battery power, you can take off one of the terminals on the car battery and work in safety. The probe also has its own

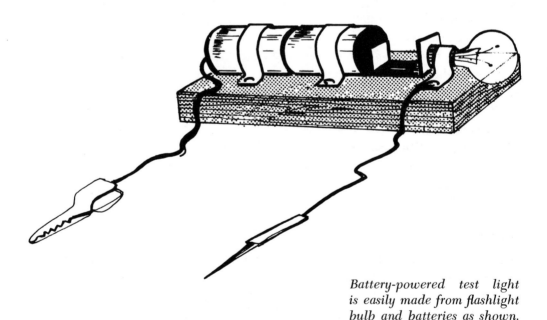

Battery-powered test light is easily made from flashlight bulb and batteries as shown.

ground so you are only testing to see whether the current-carrying wire is doing its job.

Let's suppose that you are testing a tail light which does not light up. Remove the lens, pull the bulb from its socket, and let the test probe contact the socket. Connect the grounding wire to the bumper or other ground. Your test probe should light.

If the probe does light it means that either the bulb was bad or the socket is not grounded. The ground is usually through the mounting screws so see that they're tight.

When the test probe doesn't light the only solution is to trace the wire back looking for a break. You can insert the probe into the wire at any point and ground it. When you find the point between which the probe lights and fails to light you have found the break.

Having trouble tracing the wiring? Look up the wiring diagrams in a shop manual for your car. Sometimes they have little pictorial representa-

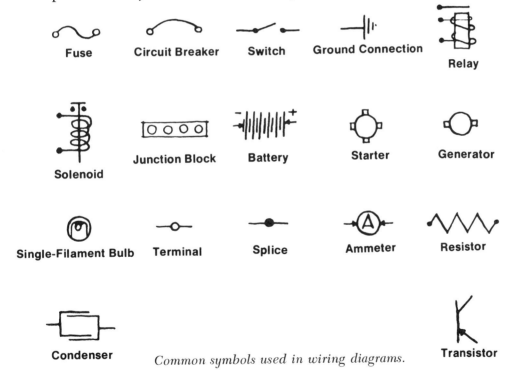

Fuse   **Circuit Breaker**   **Switch**   **Ground Connection**   **Relay**

Solenoid   **Junction Block**   **Battery**   **Starter**   **Generator**

Single-Filament Bulb   **Terminal**   **Splice**   **Ammeter**   **Resistor**

Condenser   *Common symbols used in wiring diagrams.*   **Transistor**

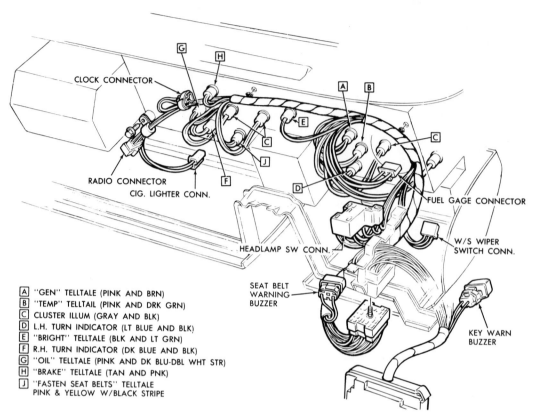

CLOCK CONNECTOR

RADIO CONNECTOR

CIG. LIGHTER CONN.

HEADLAMP SW CONN.

FUEL GAGE CONNECTOR

W/S WIPER SWITCH CONN.

SEAT BELT WARNING BUZZER

KEY WARN BUZZER

A "GEN" TELLTALE (PINK AND BRN)
B "TEMP" TELLTAIL (PINK AND DRK GRN)
C CLUSTER ILLUM (GRAY AND BLK)
D L.H. TURN INDICATOR (LT BLUE AND BLK)
E "BRIGHT" TELLTALE (BLK AND LT GRN)
F R.H. TURN INDICATOR (DK BLUE AND BLK)
G "OIL" TELLTALE (PINK AND DK BLU-DBL WHT STR)
H "BRAKE" TELLTALE (TAN AND PNK)
J "FASTEN SEAT BELTS" TELLTALE
PINK & YELLOW W/BLACK STRIPE

Color coding is an aid to locating wires. Shop manuals give diagrams and may have pictorial view as shown here. (Chevrolet)

tions of each electric device in the circuit such as relays, bulbs, fuses, resistors, and the accessories themselves, rather than making use of electrical symbols which many people don't understand. The color of each wire should be indicated on the diagram. A different wiring diagram will be given for each circuit.

Wiring can be tricky since one accessory can be wired through another and (theoretically, at least) a burnt-out bulb in a back-up light could cause the horn not to operate. Actually, wiring is usually pretty logical and the accessories wired together are related (power seats and power windows, for instance).

You can repair or replace a faulty wire. It should be replaced with wiring of the same thickness (gauge) and run through the same path. If the break was caused by rubbing, it may be wise to attach the wire to a stable member at this point to prevent future grief.

Splicing a broken wire involves more than just intertwining the broken ends and taping. The ends must be thoroughly intertwined and then soldered with a rosin-core solder to make a good electrical connection. Plastic tape should be neatly wrapped around the juncture and a final covering of liquid rubber applied. Then you'll have a connection as strong or stronger than the original one.

For those who don't solder, there's an easier way. Solderless terminals and tube-like wire joiners are sold in electrical and auto supply stores. To splice a wire you just strip some insulation off both ends and insert them into the plastic covered terminal tube. Terminal pliers are used to crimp down the metal connectors within the tube and make a firm connection. Solderless terminals of various sorts are available; keep an assortment in your tool kit at all times.

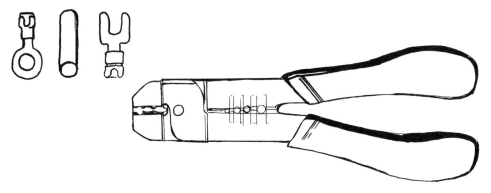

Solderless terminals and crimping pliers.

SWITCHES

When you press your foot on the brake pedal and the stoplights in back of the car go on, you have activated a switch whether you know it or not. The switch is in the housing of the hanging brake pedal. (Some are in unit with the master cylinder.)

A simple mechanical switch has contacts which come together and complete an electrical circuit when you push a button or move a toggle lever. A

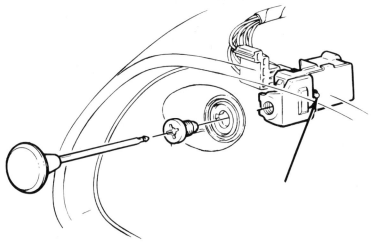

Headlight switch. The arrow indicates the release button. (Chevrolet)

relay is a switch activated by electricity. Current coming into the relay sets up a magnetic field which causes contacts to come together.

Many switches are not made to be repaired. They are simply replaced with a new unit. You can test a switch with a voltmeter by connecting a power source (battery) to one side and any accessory which requires electricity to operate (electrical load) to another. The voltmeter measures the resistance (voltage drop) in the switch. If the switch produces a resistance of more than a fraction of a volt in the voltage load it was designed to carry, the switch is bad.

An easier way is to simply replace the switch with a new unit if you suspect it is faulty. Most switches are easy to replace but a few can present problems.

The headlight switch, which also controls parking lights and dash lights (with a built-in rheostat), generally has a button somewhere on it which must be pressed to release the knob. Once the knob is pulled off, the switch can be taken out by unscrewing the chrome ring which holds it in. On some cars it becomes a bit more difficult because you must remove a part of the dash to get at the back of the switch. You might even have to take out the ashtray and work through the ashtray opening.

Ignition switches also come out and can be replaced with new units.

Many ignition switches have a tiny opening in front through which you must stick a pin before the switch can be removed. The replacement switch may come without a lock cylinder; the original lock cylinder fits into it.

At one time it was a favorite trick of car thieves to jump the ignition by bridging the terminals on back of the switch. Most modern ignition switches are constructed so this can't be done. Now thieves must yank out the switch with a slide-hammer device and connect the wires to their own ignition control gadget before driving away with your car.

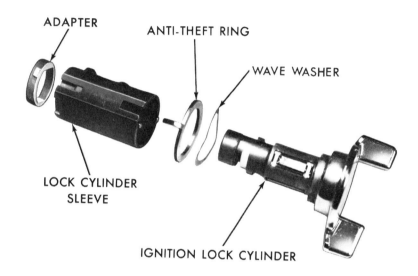

Components of ignition switch. (Chevrolet)

Dimmer switches can be removed by loosening a nut underneath the car which holds the switch on. Before you do this, however, check to see that the multiple connector for the three terminal wires to the switch has not come loose and that the wires are tight on their terminals.

Back-up light switches work off the transmission linkage. The turn signal switch is usually in the steering wheel housing and you must take off the steering wheel, using a special puller, before you can get at it. The turn signal switch is cancelled by a ratcheting mechanism which works off the steering action.

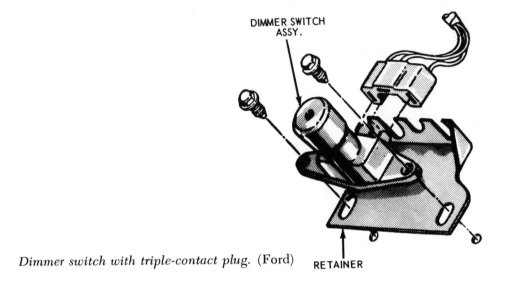

DIMMER SWITCH
ASSY.

RETAINER

Dimmer switch with triple-contact plug. (Ford)

Electric windshield wipers, blowers, power seats and windows, horns, and other electrically operated devices on the modern car all have their own switches and relays which sometimes fail and must be replaced.

Operation of mechanical stoplight switch. Contact closes when driver depresses brake pedal. (Ford)

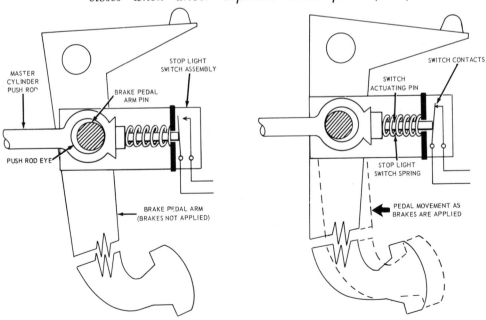

MASTER
CYLINDER
PUSH ROD

BRAKE PEDAL
ARM PIN

STOP LIGHT
SWITCH ASSEMBLY

PUSH ROD EYE

BRAKE PEDAL ARM
(BRAKES NOT APPLIED)

SWITCH CONTACTS

SWITCH
ACTUATING PIN

STOP LIGHT
SWITCH SPRING

PEDAL MOVEMENT AS
BRAKES ARE APPLIED

HORNS

Horns are electrically operated and generally sit behind the grille. You can buy fancy horns with a trumpet shape which operate from compressed air, but common electric horns usually are disc-shaped and have a front grille. They resemble miniature radio speakers.

There is no routine maintenance required on a horn so you probably won't bother with it until it becomes weak or fails altogether.

Pushing the horn button and getting no honk can mean bad wiring, a blown fuse, a faulty relay, or trouble with the horn itself. A weak note generally means wear in the horn mechanism which is not letting the correct amount of current through.

Trace the wiring back from the horn(s) to a little box which is the relay. You'll remember that relays are electromagnetic switches which close a circuit when current is sent through them. Have someone operate the horn while you listen for a click indicating the relay is closing properly.

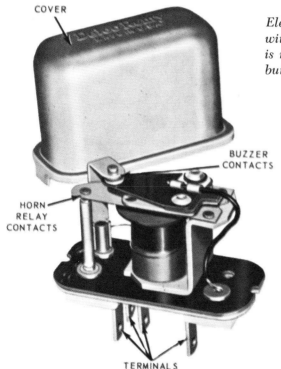

COVER

BUZZER CONTACTS

HORN RELAY CONTACTS

TERMINALS

Electrical relay (shown) is used with some horn systems. If there is no relay, contacts are at horn button. (Pontiac)

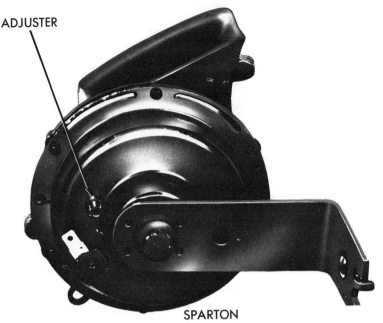

ADJUSTER

SPARTON

Horn with external adjuster. When there is no external horn adjusting nut, internal adjuster is reached by bending metal tabs to separate halves of horn body. (Dodge)

No click? Troubleshoot the wiring, check the fuse, replace the relay if necessary. (Incidentally, don't worry if you can't find the relay at all. Some cars have electrical contacts in the horn button to serve in place of the relay.)

A functioning relay indicates the trouble is in the horn. Pull the horn out to work on it if necessary. Then use pliers to bend the cover tabs and pull the cover off. In the center of the horn mechanism is an adjusting nut or screw, and a lock nut. Loosen the lock nut a bit enabling you to turn the adjusting nut while someone else sounds the horn. Turning the nut one way will increase volume while the other way volume will go down. Adjust to the right volume and then re-tighten the lock nut. Horns which will not adjust to sound loud enough need to be replaced.

DASHBOARD GAUGES

Dashboard gauges were mechanically operated at one time, now they are mostly electric. The fuel level gauge, oil pressure gauge, water temperature gauge, and ammeter receive their information in electrical impulses and the indicator needles are moved correspondingly.

The indicator needles are actuated by the familiar bi-metallic principle. As a bi-metallic coil receives more current flow it is heated and expands. The needle is pushed up the dial.

The units which send these electrical impulses vary. For fuel level a float within the gas tank moves up or down and contacts a variable resistor. The resistor controls the amount of current sent to the gauge.

The water-temperature sending unit registers heat, naturally, and uses a bi-metallic coil to do so. The oil pressure sending unit registers pressure on a flexible diaphragm.

The most familiar problems of gauge operation are not caused by either the sending unit or the gauge itself. When all gauges fail to register, a blown fuse is probably the culprit. Possibly an ignition wire has come loose. When all gauges continually register maximum it is usually the instrument voltage regulator at fault. This little device is employed on Ford and Chrysler Corporation cars to limit the voltage going to the dash gauges. It is in the electrical circuit between the ignition switch and the gauges and is usually located behind the dash. On some Chrysler products it is an integral part of the fuel gauge.

Even though the regulator is behind the dash, you can test its operation without taking the dashboard off. Just remove the voltage-transmitting wire from one of the sending units, such as the oil pressure sending unit or water temperature sensor (both usually set into the engine block). Borrow a 12-volt light bulb from a convenient socket on the car. Touch its base to the engine block (or other good ground) and turn on the ignition. Touch the wire from the sending unit to the bulb and it should flicker if the instrument voltage regulator is working. It won't light brightly since the regulator limits the voltage to less than 6 volts.

You can replace the regulator yourself, particularly if you have one of the quick disconnect dashboards as used by the Pinto and Vega and some other new cars.

The problem of all gauges reading maximum could be caused by a bad ground connection. Since most instrument clusters are grounded by the mounting screws, tighten them first and see if the trouble disappears. If not, use a jumper wire between the cluster and a ground to see if the in-

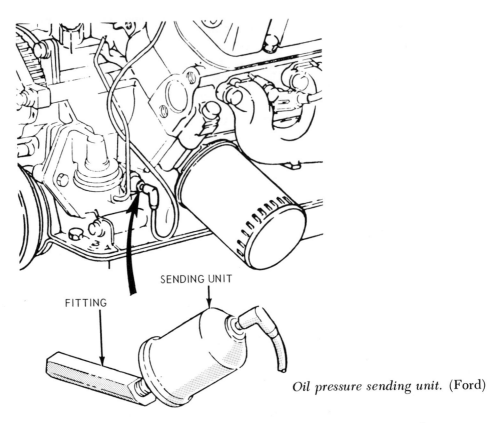

FITTING

SENDING UNIT

Oil pressure sending unit. (Ford)

struments will work. Should this solve the problem you can always run your own ground wire to save the trouble of tracking down the faulty one.

When individual gauges go bad, the cure is to locate the particular sending unit and replace it. Consult the factory manual for sending unit locations on your car. The one that will give you the most trouble is the gasoline level sending unit in the gas tank since the gas tank will have to come off. When replacement of the sending unit won't effect a cure, you'll have to troubleshoot the wiring.

WINDSHIELD WIPERS

Most wipers are electrically operated these days except those on some American Motors products which operate on engine vacuum. Older cars frequently have vacuum wipers.

Wipers which are operated by vacuum taken directly from the manifold will operate more slowly or almost stop when the engine is pulling hard (such as going up hills) and manifold vacuum drops. More recent designs use a vacuum booster which is frequently built in unit with the fuel pump. When the wipers fail and the hoses and linkage seem all right, the whole fuel pump may have to be rebuilt or replaced.

Electric wipers are run by a motor located behind the dash or on the firewall. The wiring circuit goes through the ignition switch and the wiper switch. There is usually a circuit breaker, which may be in unit with the wiper switch. The ground is generally in the wiper motor mounts.

When electric motors fail to operate, the first step is to troubleshoot the wiring and see that the motor ground screws are tight; then test or replace the switch. If the wipers work normally but fail to stop, a separate switch known as the wiper motor park switch is probably at fault. This is the switch that ensures that the wipers return to park position when you shut them off. It is wired through the control switch.

Between the wiper motor and the arm which holds the blade is a linkage of cranks and pivots called the "wiper transmission." A single blade that fails to function or operates unevenly is indicative of a faulty wiper transmission. A replacement wiper transmission is sold as a complete unit; try a wrecking yard for a lower price.

Components of electric windshield wiper motor. (Ford)

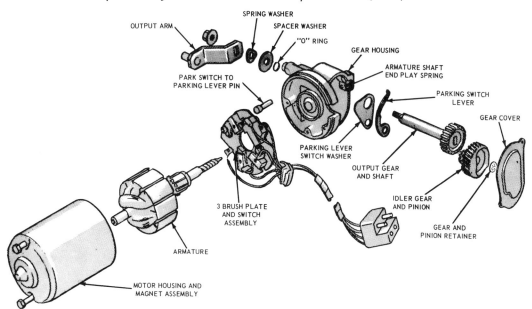

When neither the wiring or switches seem at fault, suspect the motor itself. Fairly frequently, light tapping with a screwdriver handle or similar object will dislodge a stuck brush and get the motor operating again—for a while. It is probably this characteristic of electric motors which has given rise to the idea that balky motors always respond to a housewife's kick—it only works sometimes.

If you can take a wiper motor out yourself it is often a good idea to bring it to a shop specializing in the repair of small electric motors rather than an automotive garage. Too many garages will just put in a whole new motor rather than taking the time to repair faulty parts.

Needless to say, any work involving the electrical system should be done with one battery cable disconnected and placed where it will not contact a ground.

On cars with wipers which sink into a well beneath the windshield, the wipers must be taken out of park position before work is done on the motor.

Trico-type wiper blade is removed by depressing tab. (Ford)

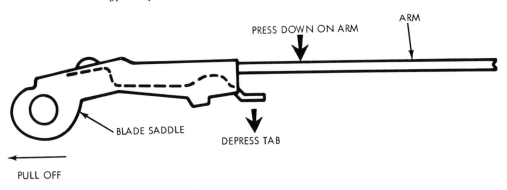

Anco bayonet blade is removed by pressing release button. (Ford)

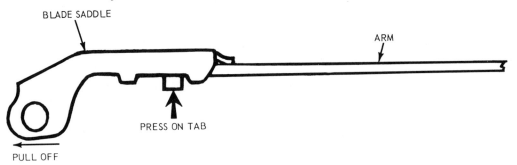

Some cars are designed so it is easy to get at the windshield wiper switches and motor, while others require removing the dash, speedometer cable, and other intervening parts. A service manual and wiring diagram are a great help.

Incidentally, not all wipers are operated by vacuum or by electricity and gearing. On 1965–69 Lincoln Continentals there are hydraulically operated wipers using fluid from the power steering system. There is an adjustable cable to a valve control on the electric motor and the hydraulic system can be checked for leaks and fluid level.

Wipers that clunk against the sides or bottom of the windshield molding can be adjusted. Pull the wiper arm away from the windshield on its pivot and remove it using a screwdriver to slide the cap of the arm off the serrated shaft. With the arm off, turn on the wipers and then turn them off so that they go into park position. (Be careful not to let the arm stub contact the windshield and scratch it.) Re-install the arm so that the tip slightly touches the side of the windshield molding.

Always wet the windshield before testing the wipers for correct operation.

WINDSHIELD WASHERS

Some windshield washers have a mechanical pump operated when you press a button on the floor board. Others have an electric pump in the fluid reservoir or next to it. A third type has a pump which is in unit with the windshield wiper motor. Operating the washer causes the wiper blades to make a few strokes.

The washer reservoir, usually made of plastic and located on the firewall, should be kept filled with windshield washer anti-freeze available at most accessory stores and many supermarkets.

The hoses and nozzles are most often at fault when washers fail to operate. You can remove the hoses and blow through them to check for obstructions. Rusty nozzles can be cleaned with a small drill bit.

Mechanical pumps are fairly inexpensive and easy to replace. Electric pumps can be serviced by an electric motor repair shop. When an electric pump is beyond fixing, look for a replacement at a wrecking yard. New pumps are expensive and rebuilts are not usually available.

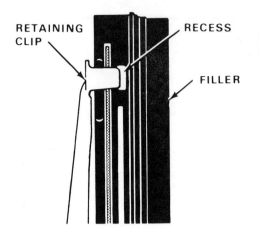

RETAINING CLIP   RECESS

FILLER

To remove wiper filler: squeeze short end of filler and twist it out of retaining clip at both sides. (Volkswagen)

STEEL STRIP   STEEL STRIP

RETAINING CLIP

FILLER

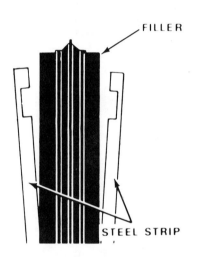

FILLER

STEEL STRIP

The second step is to move filler to side and slide steel strips out. Then slide filler out of remaining clips. (Volkswagen)

To replace wiper filler: be sure notches in strips face new filler. Place steel strips in grooves. Starting at open end, slide filler into retaining clips. Make sure clips ride in recess of filler. (Volkswagen)

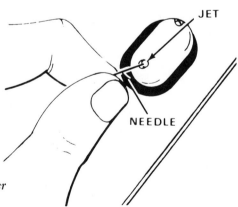

JET

NEEDLE

Thin needle can be used to unplug washer jets or reorient spray. (Volkswagen)

RADIOS AND STATIC

The repair of car radios, now manufactured with transistors and printed circuits, is beyond the scope of most auto mechanics. You may not want to mess with the innards of a faulty radio either, but there are some simple troubleshooting steps which can clear up many of the more common problems.

When the radio goes dead, always check for a blown fuse. On factory installations the fuse for the radio circuit can be found in the fuse box, neatly labeled. Aftermarket installations often utilize an in-line fuse which can be found by examining the wires at the rear until you locate the plastic fuse casing. In either case, replacing the fuse with a new one of the same amperage may restore your music power.

Also look for loose or broken wiring, or a bad ground connection. The ground may be through one of the screws holding the radio in place, so check them all for tightness.

Weak signals could be due to a faulty antenna connection. Be sure the antenna is mounted securely (the mount screws are the ground) and that the cable from the antenna is fastened tightly to the radio by a connector plug. Replace corroded connectors.

Beneath the mount cover plate of the antenna is a plastic insulator. If it is dirty or cracked, it can short out the antenna. So take the plate off and clean or replace the insulator.

There is also an antenna "trimmer" screw on the radio, probably under the tuning knob in front or next to the cable socket at the rear. A radio must be tuned to its antenna in order to play properly, and sometimes the screw gets out of adjustment. To make sure the trim is set correctly, just tune in on a distant station and turn the screw to the point where the signal comes in strongest.

Of course, when the antenna is broken off by neighborhood vandals, you will not get adequate reception until it is replaced. A piece of coat hanger wire can do the job in a pinch, at least until you have a chance to stop in at the nearest supplier of auto parts and buy an antenna that fits over the old stub. It tightens in place with a setscrew.

Static is another problem you should be able to clear up on your own. The first step is to note whether the static is present only when the car is moving, when the car is stopped with the engine idling, or only when a particular electrical accessory is turned on.

A popping noise heard only when the engine is running can usually be cured by proper ignition suppression. Suppressor kits are sold at auto gadget stores and consist of clip-on suppressors for the spark plugs plus shielding for the distributor and coil. You may have to replace the spark plug wires if they have frayed or cracked insulation. Resistance plug wires, rather than hot-rodder's metallic-core wiring, are necessary.

Popping or crackling noises, which occur only when the car is moving at a fair clip, are usually caused by wheel or tire static. Inside the dust covers on the front wheel hubs there should be thin metal static collectors. If they are missing or defective, try installing static collector springs. These cost a few cents apiece and are small wire coils which are pushed on the ends of the wheel spindles just beneath the dust covers.

Tire static can be cured by an antistatic powder which is blown in through the tire valve with a special tool. Many radio repair shops are equipped to do this job.

Whining or buzzing noises which rise in pitch as the engine is speeded up indicate interference from the alternator or generator. Sometimes the problem can be alleviated by replacing worn brushes which cause sparking. Otherwise your best bet is to install a tuned-trap filter (purchased from an electronic supply store) on the armature terminal of a generator or battery terminal of an alternator.

In cases where the radio starts whining and spitting only when a particular accessory (such as the windshield wipers) is turned on, the solution is to install a tuned-trap filter, or another little gizmo called a coaxial capacitator, to eliminate interference. Consult a radio repairman, or one of your friends who plays around with ham radios and has the reputation for being an electronics whiz, for advice.

When all else fails, remove the radio and bring it in to a repair shop. Most radios can be readily taken out by detaching the wires at the rear (be sure you label them first), pulling off the control knobs at the front and unfastening the attaching sleeve screws behind them, and removing the support bracket(s). Some radios are installed in two parts, or are difficult to get at, in which case it is best to consult the manufacturer's shop manual for a step-by-step procedure.

STARTER SOLENOIDS

The starter solenoid is a special kind of electromagnetic switch or relay. When it is activated by turning the ignition switch to the start position, an iron "plunger" within the solenoid is moved with considerable force and not only makes the electrical contact but also moves the starter pinion gear into engagement with the ring gear on the flywheel. The flywheel is connected to the engine's crankshaft so the starter motor can make the crankshaft revolve and start the cycle which activates the engine.

The three major car manufacturers in America use different types of solenoids. Ford products are fitted with a starter which does not need a true solenoid, so a simple relay mounted on the sidewall or firewall is used.

Chrysler products use both a relay mounted on the firewall and a solenoid on the starter. GM cars just have a solenoid on the starter.

You can find out if a solenoid (or relay) is defective by using a jumper cable to bridge the terminals which connect the solenoid to the starter and

Exploded view of solenoid. (Chevrolet)

SOLENOID BODY

TO HOLD-IN COIL

SWITCH TERMINAL

CONTACT RINGS

FIBER WASHER

CONTACT FINGER

BATTERY TERMINAL

TO PULL-IN COIL

PLUNGER

MOTOR CONNECTOR STRAP TERMINAL

END COVER

to the battery. When this is done with the ignition on, the car will start if the starting problem was a defective starter switch. Some relays (such as those on some Ford products) have a button on bottom to manually activate them when they won't operate normally.

Ford starter relays are sealed (with the exception of some older models) and can't be repaired. Solenoids mounted on the starter can be fixed, although in some cases you must remove the whole starter assembly to work on them.

Most often the problem with a solenoid is worn contacts. There are usually fixed contacts attached to the cover plate and a brass contact disc which moves to touch them. These contacts can be filed smooth if they are pitted or covered with deposits.

Another way to repair the contacts is to simply snap out the contact disc and turn it over. The underside of the disc provides a new contact face. One of the fixed contacts (the one which connects to the solenoid's battery terminal) can be rotated by loosening its retaining nut. Turn it so that the part which is not pitted comes in contact with the moveable disc. When the contacts are badly burned or too far gone to be reversed, you may be able to buy a solenoid rebuild kit and install new contacts, thus saving a few dollars.

Sometimes a solenoid failure can be caused by a sticking plunger or bad windings. See that the plunger works smoothly when you put the solenoid back together. If the windings are defective you will need a whole new unit.

STARTERS

The starter consists of both the motor itself and the drive mechanism which contacts the flywheel gear.

When the drive pinion gets stuck on the flywheel gear it may be possible to get it off without disassembling the drive unit. On a manual transmission car this is done by putting the car in second gear and rocking it backwards and forwards until the pinion snaps out of engagement. This can't be done on an automatic transmission car, but you can try banging on the drive mechanism with a rubber hammer (or a block of wood) until the pinion frees itself.

Any further repairs will require removal of the whole starter mechanism —not an easy job on some V-8 cars where you must work from underneath the car in limited access space. You may even have to remove the exhaust pipe.

Assuming that you can get at the starter easily, it is not difficult to remove it. After taking off the terminal cables (label or color-code them so you remember which goes where) unbolt the starter from the flywheel housing and lift or lower it out.

You are faced with a choice of replacing the whole starter (a used or rebuilt unit will help your budget) or opening it up and trying to replace the defective parts.

Many service garages will not bother trying to fix a starter—they claim that the labor charges will probably exceed the cost of a rebuilt unit.

If a hung-up starter drive was your problem, and if the starter drive can be removed without taking the whole starter apart, it pays to install a new drive unit only.

In cases where the starter itself is defective, or the internals must be removed to replace the drive mechanism, it probably makes sense to replace

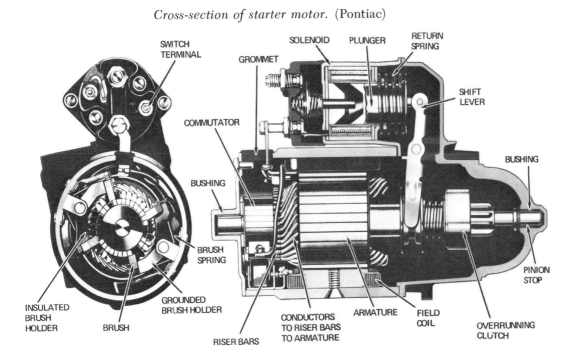

Cross-section of starter motor. (Pontiac)

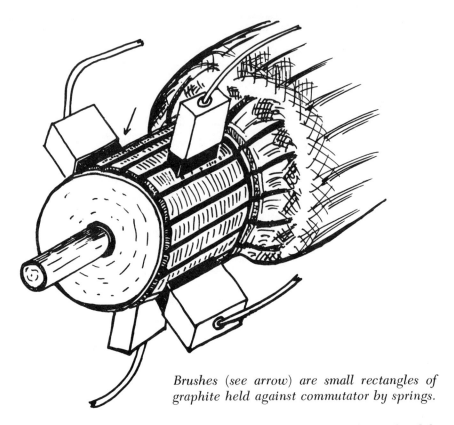

Brushes (see arrow) are small rectangles of graphite held against commutator by springs.

the whole unit. Starters are put together with snap-rings which can be difficult to remove and replace even if you have the proper snap-ring pliers with a set of tips which fit. Special tools are required to test the armature and turn a commutator with excessive runout. Much as it may gall you to discard (or trade in) a starter which may have only a single defective part, it usually makes good economic sense.

One repair you may be able to make on a faulty starter is to replace worn brushes. Before we go into this job, however, a short discussion of how electric motors work is indicated.

A starter converts electrical energy to mechanical energy and therefore resembles the electric motors in kitchen appliances, power tools, and many other electrically powered pieces of machinery.

An electrical current flowing through a wire produces a magnetic field around the wire. In the starter the electricity goes through coils wrapped around pole shoes which are thus converted to powerful magnets. On the shaft of the starter is an armature, made of soft iron, which is also wrapped with electrical windings so that it becomes magnetized.

When you were a kid, you probably played with magnets and noticed how like poles repel each other and opposite poles attract. You could use this principle to move one magnet around with the other, even though they never touched.

Electrical current makes the armature rotate by this attraction/repulsion principle. The armature is attracted by one pole and repelled by another, so it makes part of a revolution. Then the magnetism of the armature is reversed at a critical point, and the armature is repelled by the first pole and attracted by the second, so it continues to revolve.

The reversal of the magnetism of the armature is accomplished by the commutator, consisting of segments of brass connected to the armature windings. The brushes ride against the commutator and provide current to the windings. The motion of the commutator, revolving on the shaft, cuts off current from one part of the armature's windings and feeds current to another.

An analogy might be with a donkey and a carrot, except that there are two carrots, in fixed positions, and the donkey is chasing around a circle after them. Since the carrot stays in one place, the donkey could reach it. The trouble is that as soon as the donkey approaches one carrot, he changes his mind and goes on to the next. Thus, he is continually chasing around in a circle and gets neither carrot.

The armature never reaches its goal, but it does do the work of keeping the shaft revolving and thus producing mechanical energy.

The brushes, small rectangles made of a soft graphite material, are held against the commutator by springs. Information on replacing brushes is covered as an emergency procedure in the next section, dealing with generators.

You can run a starter while it is out of the car by using jumper cables. Run one cable from the positive battery terminal to the main starter terminal and the other from the negative terminal to the starter frame. This is a good way of telling whether or not a starter you are about to buy in a junkyard actually functions.

There is a good possibility that the problem with your starter is due to

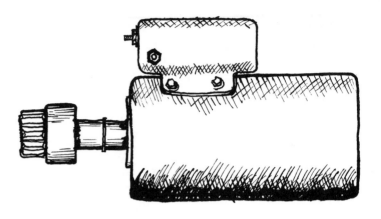

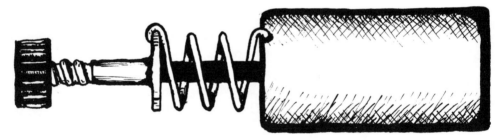

Two types of starter drive mechanisms. Overrunning clutch drive (top) has sole-noid directly on starter. Bendix drive (bottom), as used on Ford products, has remote relay.

worn bearings or bushings, since these are the parts that usually go first. If the armature will not revolve freely when turned by hand, or the shaft is loose enough to have noticeable play, plain old wear is the root of your problem. You just might be able to find new parts at an auto supply house and install them yourself. A better bet would probably be to deliver your starter to a shop which specializes in the repair of electric motors and see if they can do the job for you.

GENERATORS, ALTERNATORS, VOLTAGE REGULATORS

The generator works like an electric motor in reverse. It converts mechanical energy (the rotation produced by the engine-driven fan belt driving the generator pulley) to electrical energy needed to keep the battery charged and run electric devices on the car.

The basic principle behind generators is that an electrical conductor produces current when it crosses a magnetic field. The armature is the conductor, the field poles draw just enough electricity to produce the magnetic field, and the current produced is picked off the commutator by the brushes and goes to a terminal from which it is sent to the voltage regulator.

The alternator has replaced the generator on newer cars because it can put out more current and even keep the battery charged while the car is idling. Generators are shaped like a cylinder, while alternators have a shape like a fat pancake and are finned or vented.

End view of alternator.

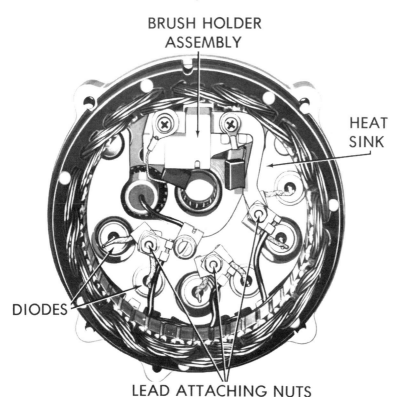

BRUSH HOLDER
ASSEMBLY

HEAT
SINK

DIODES

LEAD ATTACHING NUTS

The alternator also produces current by having an electrical conductor intersecting a magnetic field. However the conductor is stationary and the field-producing element is the one which revolves. The rotating element is called the *rotor* and it revolves inside a ring of coils, known as a *stator*. The current goes to brass *slip rings* rather than to a commutator, and is picked up by brushes rubbing against these rings.

The reason that alternators can produce more current is that they generate alternating current (AC) rather than direct current (DC). The alternator is frequently called an AC generator.

The generator only uses current of one polarity while the alternator picks up current of both polarities and the current is constantly changing direction.

Since the electrical devices on a car are designed for direct current, the alternator has rectifying *diodes* which convert AC to DC. The diode is a type of transistor that permits current to flow through it in one direction only.

Voltage regulators used with generators look like little black boxes and contain three relays, each having a separate function. The *cut-out relay* makes and breaks the connection between the battery and generator. When the generator is revolving fast enough to charge the battery, the regulator makes the connection. Whenever the generator slows down and is not producing enough power to charge the battery, the circuit is broken. Sometimes the cut-out relay is called a circuit breaker.

The *current limiter* relay puts a ceiling on the amount of current the generator produces. Since current production varies according to how fast the engine is turning over (which controls the speed of the fan belt and therefore the rotation of the armature within the generator), the generator output constantly varies. The current limiter keeps current output to a safe level.

The third relay is a *voltage limiter*. Voltage is the pressure that forces electrical current (amperage) through transmission lines. A useful analogy is with a water faucet. The amount of water flowing through the faucet is equivalent to amperage in an electric line while the water pressure corresponds to voltage.

If the cut-out relay is faulty, the battery can discharge itself through the generator. A bad current-limiter relay would allow the generator to at-

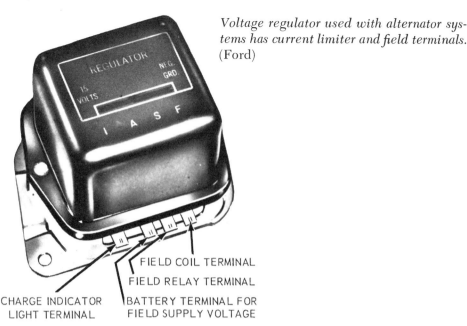

Voltage regulator used with alternator systems has current limiter and field terminals. (Ford)

FIELD COIL TERMINAL

FIELD RELAY TERMINAL

CHARGE INDICATOR
LIGHT TERMINAL

BATTERY TERMINAL FOR
FIELD SUPPLY VOLTAGE

tempt to produce too much current and burn itself out. A malfunctioning voltage-limiter would send too much voltage to the electrical devices on a car and cause them to burn out. More likely, fuses would blow.

Voltage regulators used on alternator systems need only the voltage-limiter relay. Those on Chrysler products usually have only this single relay while other cars frequently have a field relay used to connect the alternator to battery current as soon as the ignition is turned on. On some cars the regulator is in unit with the alternator.

The charging system (generator or alternator and voltage regulator) requires very minimal maintenance. Generators usually have a cup which is periodically filled with a few squirts of oil. Alternators and voltage regulators require no lubrication.

Troubleshooting is another story. When the red light on your dash comes on or the ammeter indicates that the battery is discharging, better get out and check the charging system.

First look at the fan belt to be sure it is not slipping and failing to keep the generator or alternator running at adequate speed. Make sure all the connections at the generator or alternator and voltage regulator are clean and tight.

Now let's see how to check out a generator system. The first step is to find out whether the generator or the regulator is causing the problem. This is done by using a jumper wire to cut the regulator out of the circuit.

Shut off the engine and turn off all electrical accessories. Then look at the two big terminals on the generator. One should be marked "A" and the other marked "F." Both the armature and field terminals are connected to the voltage regulator. The three relay terminals on the regulator are usually marked "ARM," "FLD," and "BAT." The armature and field terminals on the regulator are hooked up to the corresponding terminals on the generator. The battery terminal is hooked up to the battery circuit and generally goes to the ammeter on the dash and then through the ignition switch.

On most GM and Chrysler products the generator is grounded externally through the regulator. All you have to do to cut the regulator out of the circuit is remove the field wire at the regulator, then use your jumper wire to connect the generator field terminal to a good ground. On Fords the generator is internally grounded so the jumper wire is connected from the generator's armature ("A") terminal to the field terminal on the regulator.

With the jumper wire in place, start the engine and rev it up a bit to see if the generator is now putting out a charge. (The warning light will go off or the ammeter needle will swing to the charge zone.) No charge means the trouble is in the generator itself. If there is a charge then the regulator is the problem.

Don't run the car with your jumper wire in place since the unregulated generator could put too much voltage through an electric accessory and ruin it. Even with everything electrical turned off it could harm the battery. Instead try a quick fix on the generator or regulator.

Tap the generator with a screwdriver handle on the end away from the fan belt. This may dislodge a sticky brush and get things perking again. Brushes are the main cause of generator trouble.

The generator may have a metal band which snaps open to allow inspection of the brushes. If so, open it and use a screwdriver to press the brushes down against the commutator. If the system now shows a charge, wedge wood or paper into the brush springs to hold the brushes in contact temporarily until you get to a service station.

When replacing brushes the commutator must be turned down on a lathe, so have the job done at an electric motor repair shop.

A problem in the regulator can also frequently be fixed temporarily by tapping it with a screwdriver handle. This will jar a sticking cut-out loose.

If the above procedure doesn't work, remove the regulator cover and find the cut-out relay. This is the one marked "BAT" and its contact points are held open rather than closed by spring tension. Disconnect a battery cable and use a strip of fine sandpaper or a nail file to clean the cut-out contacts. Hold them together and pass the abrasive through. Run a clean strip of rag through the contacts to remove any filings. Now put the cover back and see if the charging system is revitalized.

On an alternator system you cannot use jumper wires to cut out the regulator. Grounding the field circuit between the alternator and regulator will harm the regulator. Grounding any terminals or shorting across them is dangerous.

Also be cautious when getting a battery boost or giving one. Even a momentary wrong connection between a positive and negative terminal can knock out your system. When getting your battery quick-charged at a service station, ask if the charger has an alternator protector built in. If not, you will have to disconnect the battery cables while charging it.

Alternators do have brushes, some of which are easily accessible without opening up the unit itself, so replacing the brushes is one troubleshooting procedure you can try.

An alternator is easy to remove from the car. Just loosen the alternator bracket bolt so you can swing the alternator in toward the engine and loosen the fan belt, then loosen the other bolts holding the alternator in place.

When you have had the alternator tested and parts replaced, or perhaps traded it in on a rebuilt model, you can install it in place again by reversing the steps you used to take it off. This is not true of a generator, which must be polarized when installed.

The generator is polarized by installing it in place and hooking up all the wires. Then, before starting the engine or even turning on the ignition, connect a jumper wire to the battery terminal of the regulator and momentarily touch the other end to the armature terminal of the regulator. That's all it takes.

If your car does not have an ammeter, monitoring the charging system output, it's a good idea to install one. It is one of the most reasonably priced instruments and the easiest to hook up. Then keep an eye out for a system which is discharging when it should be charging and avoid those electrical emergencies which always seem to culminate at night on a lonely road.

WATERPROOFING THE IGNITION SYSTEM

As long as the electrical system of a car is shielded from water, the carb is getting a supply of air to mix with the gas, and the exhaust fumes have some place to go, there's no reason why a car can't be run underwater.

The Army has converted military jeeps to ford deep rivers with a kit which includes waterproof casings for the distributor, coil, and plugs, along with an extension to run the exhaust straight upward like a snorkel tube.

In theory, at least, you could seal up the doors and other openings on your car for proper flotation, install a complete waterproofing kit, hang an outboard engine on the back, and become the first man to cross the Atlantic in a 1967 Ford Galaxie with air-conditioning, and Ray Charles blasting out of your tape deck all the way.

Of course, there are waves . . .

A more realistic goal is to make the ignition system water resistant enough so that it will not drown out as you splash through deep puddles. Also you won't have as much trouble starting on damp days.

Waterproofing kits are available from a few car manufacturers (Mercedes, for one) and a number of electrical accessory suppliers. A kit usually consists of a special sealed distributor, coil cover, and nipples for a tight seal around the plug ends. The kits are easy to install but relatively expensive.

You can do the job yourself with a tube of GE Silicone Seal (clear variety) available at just about any hardware store. The first step is to use a rag and a little solvent to remove all the grease and crud from the exterior of the distributor, coil, and ignition wires.

Applying the Silicone Seal with the tip of your finger, use it to seal around all the little caps which go on the spark plugs and the corresponding nipples which cover the plug wires at the distributor cap end. Press all connections on tightly and wipe away the excess sealant. Then remove the distributor cap and apply sealer around the upper and lower mating surfaces before replacing it.

Also seal the nipples which hold the secondary (thick) wire to the top of the distributor and coil. Put a coating on the other coil terminals. Then spray all the ignition wires with an aerosol water-resistant coating (commercially called ignition spray or ignition seal).

The car is not totally waterproof after this treatment since there are other places (solenoid, alternator, and battery, for instance) where water could cause a dead short and stop the car. However, coating the ignition parts with Silicone Seal will protect the areas of the electrical system which are by far the most vulnerable to the effects of a wetting.

One reason you can't waterproof the battery, alternator, and other parts with sealant is that they need to be vented. Even a distributor sealed with silicone will eventually suffer from internal water condensation. If you want to keep the distributor sealed for a long time and still prevent internal corrosion, the solution is to drill a vent hole in the side of the cap, fit it with a thin piece of rubber tubing, and run the other end of the tube to your air filter canister so that dirt does not get into the distributor.

As you might expect, it will be necessary to cut through the sealant come tune-up time. Only you can decide whether it's worthwhile.

5

Chassis, Power Train, Body

● If you have ever ridden in a vehicle with solid wheels, such as an old-time hay wagon, you may very well have come to the conclusion that mankind's greatest invention was not the wheel but the tire.

The tires are the real shock absorbers on a car; they apply the power to the road, turn whenever you move the steering wheel, stop when you put on the brakes. The whole sophisticated mechanism of the modern car is directed toward controlling those four small patches of rubber which contact the road.

There was a time when selecting a new set of tires for your car was a relatively simple matter. The tire size was indicated by a designation such as 700-14, meaning that the tire was seven inches wide (or close to it—the nominal designation is not exact) and made to fit a wheel fourteen inches in diameter. Tires with four-ply construction were standard while six or more plies indicated a tire for heavy-duty use. The tires, once rubber with fabric plies, became artificial rubber with nylon or rayon plies, and are now made from additional materials.

The days of simple tires are past. Now, when Mrs. Wallaby heads down to the local tire emporium for a new set of treads, the following exchange is likely to take place:

Mrs. W.: I'd like to buy four new tires for my car, please.

Tire Salesman: What kind of tires would you like, Madam?

Mrs. W.: Well, I'd like some fairly good tires, not too expensive, but durable.

Salesman: Bias-ply, belted, or radial?

Mrs. W.: Whichever are best, I suppose.

Salesman: Radials are the finest tires. Let's see, your car would take a 185R14 tire or you could go up to an ER70-14.

Mrs. W.: But my husband told me to be sure to get size 700-14.

Salesman: That would be the size in an older conventional tire, Madam. However that designation is now 735-14. Of course, we could also give you an excellent belted-bias tire in size E78-14.

Mrs. W.: It's all so confusing. Maybe you'd just better give me a radial tire in the first size you mentioned. What would it cost?

Salesman: About $60 apiece, Ma'am.

Mrs. W.: Ridiculous. I won't spend more than $25 per tire. What is a good tire at that price?

Salesman: That depends on the type of tire you want. The tires I showed you were steel-belted radials. What construction material would you prefer?

Mrs. W.: Well, er . . . rubber, I suppose.

Salesman: The tread and sidewall of a tire is rubber, some sort of artificial rubber, that is. However I was referring to the cord material. Would you like a rayon, nylon, or polyester tire? Or I could give you a belted tire with a fiberglass belt.

Mrs. W.: I think I'll just take the nylon one. Is it a 4-ply?

Salesman: It's 4-ply *rated*, Ma'am. While it only has two plies, it is the equivalent of a 4-ply tire. Let me show it to you.

Mrs. W.: This tire has a brand name I never heard of. Couldn't I get a tire made by one of the major manufacturers?

Salesman: Actually this tire *is* made by a major manufacturer although we sell it under our own brand name. You'll note that this tire has the number 127 molded into it. Those numbers are the federal code for Uniroyal.

Mrs. W.: Then how come that tire over there has a whole string of letters and numbers?

Salesman: Oh, that one is the new federal code. The first letters are "AJ" which is also a designation for Uniroyal. The rest of the code tells you the plant where the tires were made, the year and week of manufacture, and other information.

Mrs. W.: Oh. Well if these tires fit and you say they're pretty good, I guess I'll take them. I don't think I want to know anything more about tires.

Salesman: Fine. Now would you like to have the tires balanced, and if so do you want them just statically balanced, or dynamically balanced?
Mrs. W.: You wouldn't happen to have an aspirin, would you?

To avoid migraine the next time you buy tires, a little advance study is necessary. Here is a brief rundown on the essentials of tire selection.

TIRE SIZE AND TYPE DESIGNATIONS

One number common to tire designations such as 700-14, E78-14, or 185R14 is the number 14—meaning that the tire is designed to fit a wheel rim 14 inches in diameter. If your car comes with 14-inch wheels, you will have to plunk down at least $40 or so for wheels 15 or 16 inches in diameter in order to fit tires larger in diameter.

Car manufacturers balance a great many considerations in selecting tire diameters and it is unwise to change unless the larger size is a factory option. Otherwise you may wind up with tires that thunk against the top of the wheel well over every bump.

While larger tires are rarely advisable, wider tires do make sense and are one of the most popular innovations of recent years. A Mustang or Firebird without fat ovals and raised white lettering on the side of the tires seems hardly authentic.

Tires which are an inch or two wider than stock can be fitted without changing wheels. Any tire dealer can tell you the limit for your car. Should you insist on a still wider tire, it will cost you a century note or two for wide "mags," which are styled wheels (i.e., designed to look good without hubcaps) sometimes made of magnesium, more often of aluminum or chromed steel. Wider painted steel wheels are also available at a somewhat lower price.

Mag wheels and ultrawide tires were born in racing circles. The wheels were designed to be light and strong. Wide tires, putting more rubber on the ground, offered better cornering capability, superior traction in acceleration and braking, and—for those who preferred racing in the boondocks over mud and sand—superior "flotation" over soft surfaces.

For general performance on the highway the real fatties have their limitations. They make a car steer like a Charles Atlas muscle-building device (power steering is usually a must), often give a harder ride, and require wheel well modifications on most cars.

Very wide tires can also be dangerous when going through puddles at a fair clip. Tires tend to "aquaplane," that is they lift and cause loss of control. Fat tires aquaplane much more than slimmer ones, a tendency which can be only somewhat reduced by tread design. When aquaplaning the only effective steering device is a rudder.

Older tire designations combined a load-carrying capacity rating with a width dimension. Thus the first part of 700-14 meant not only a nominal width of 7 inches, but also a certain weight that the tire could presumably bear. New designations recognize factors other than width which affect the load-carrying capacity of a tire. In the designation E78-14, the "E" signifies the load range and "78" the width (or rather the "aspect ratio," a term we'll deal with shortly).

Load-carrying capacity ratings on tires for conventional cars go from A (equivalent to the old 600) to L (915).

The aspect ratio is expressed as a percentage. A "78" tire is 78 per cent as high as it is wide, the height in this case being cross-sectional. Tires with an aspect ratio of 78 are considered standard, 70-series tires are wide, and the 60-series comprises the real chubbies which are more for racing than street use. The convenience of designating tire width by aspect ratio is that this proportion can stay constant and always indicate the standard widths for a particular diameter. Thus a 78-15 tire is fatter than a 78-13, but each tire is a standard width for its size.

Tires in the 70 series are often fitted as wider replacements for standard 78-series tires. They increase traction, braking ability, and cornering speed. However a 60-series tire is usually too much of a good thing. A tire this size will require special wheels and will probably be too wide to fit within the wheel wells of your car without bumping against the fenders when you steer from lock to lock. On four-wheel-drive vehicles and other cars where ultrawide tires are really needed for traction, load-carrying capacity, or "flotation" on soft surfaces, wheel wells can be modified with fiberglass fender extensions affixed with pop rivets. Or the wheel well sheet metal can be "massaged" by an experienced body and fender man.

The "R" in a tire designation means that it's a radial. On foreign tires the width is often given in millimeters while the diameter remains in inches. A

185R-14 tire has a nominal width of 185 millimeters (there are about 25½ millimeters per inch), is a radial, and is made to fit a rim 14 inches in diameter. Millimeter sizes for conventional tires run from about 135 (520) to 235 (915).

Some foreign tires are marked with the letters "S," "H," or "V." These indicate sustained speed capabilities: S to 112 mph, H to 130, and V to 165. American tires marked with the new speed-capability code mandated by the National Highway Traffic Safety Administration have much more modest limits designed to apply to the ordinary tire rather than sports-racing models. The NHTSA code is A to 85 mph, B to 95, and C to 105 mph.

The new NHTSA code also rates the tread life and traction of new tires. Tread life is rated with numerals from 1 to 6. The lowest number indicates that the tire has only 60 per cent of a standard or "average" tread life. Number 6 tires have a 200 per cent tread life, or double the average.

Traction is rated with one to three stars. A one-star tire has 10 per cent less traction than standard, while a three-star tire will be 10 per cent above standard. Since the standard is necessarily arbitrary and traction is rated only under certain conditions, these traction ratings are somewhat less than universally applicable. They are simply a starting point in dealing objectively with a very complex area.

Here's a little quiz to see whether you've been absorbing all this. On the sidewall of a tire you find the marks "5°°B" and also "205R-15." Can you interpret this? Well, the "5" means a tread life longer than the standard "3" (150 per cent longer to be exact), two stars indicate average traction, and "B" means a sustained speed capability of 95 mph. As for the size, it's 205 millimeters wide (about 7.75 inches), radial, and made to fit a wheel 15 inches in diameter. In this case you were not given the load-carrying capacity (the "A" in A78-13, for instance), but where given it does correspond to a certain weight per tire and is most useful when you are calculating whether or not you have enough capacity to pull a trailer or carry a heavy load (more about this in Chapter 11).

The tire will also have information on the type of construction, an inflation recommendation, and the manufacturer's code. Codes indicating who makes what tire are lengthy and they are frequently changing, so they cannot be covered here. If the topic interests you, send away for a publication called *Who Makes It? And Where?* available for $1 from Tire Guide, 2119

Route 110, Farmingdale, N.Y. 11735. It is published annually along with another handy booklet, the *Tire Guide* ($3), which lists standard and optional tire sizes for American and most foreign cars, plus other useful data on tires.

TIRE CONSTRUCTION

At one time the choice was tube or tubeless, nylon or rayon plies. Now most car tires are tubeless and they may be bias-ply, belted-bias, or radial in a variety of materials including polyester, fiberglass, and steel cord.

Tubeless tires are favored because they blow out with a whimper rather than a bang and because they are safer in high-speed operation. Tubes are more likely to succumb to heat build-up.

Truckers and those who operate four-wheel-drive vehicles often favor tubes because they are easier to mount and dismount from the rims and can frequently be repaired by installing a new tube or patching the old one. Under rugged conditions they are less likely to lose air and don't automatically fail when a wheel is bent.

Sometimes a tubeless tire is repaired by installing a tube. This is a temporary measure at best since tubeless tires are constructed differently and the tube is subject to internal friction when installed in one. It probably won't last long.

A radial tire has plies that run perpendicular to the tread and have a reinforcing belt (the cords in a bias-ply tire are diagonal, or at about a 45-degree angle to the tread.) The sidewall on a radial is quite flexible. Such tires keep better contact with the road and do not "squirm" as much when going around corners. The result is better handling, superior tire wear (as much as two and a half times the conventional bias-ply tire), and slightly better gas mileage. Radials cost a good deal more than conventional tires and they cannot be mixed or handling will suffer. You should buy four radials and a spare when converting.

Belted-bias tires are a compromise between conventional bias-ply tires and radials. The belt makes them stiffer, giving the advantage of radials but to a lesser extent. They can be mixed with bias-ply tires, however both tires on one axle should be of the same type. Belted tires give a somewhat stiffer ride than bias-ply tires yet are not as stiff as radials.

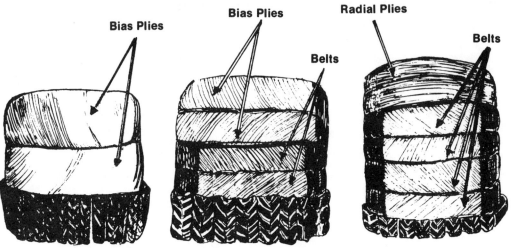

Tire construction: (left)bias-ply; (center)belted-bias; (right)radial.

The type of material used in constructing a tire does not necessarily determine the tire's quality. Rayon is traditionally used for the plies on inexpensive tires, but is also employed on deluxe radial tires with steel cord belts. Nylon is a strong material used on many quality tires, but it does have a tendency to "flat-spot." When a car with nylon tires is left standing overnight, it will go "thump-thump" as you pull away from the curb. This flat "set" is only momentarily annoying; it will disappear after a few miles. Polyester body plies and a fiberglass belt are a popular combination for belted-bias tire construction.

At one time retreads were the cheapest tires around and were often unsafe. Now there are federal standards for retreads so your chances of getting a dud are small. The trouble is that retreads meeting the new standards have become a good deal more expensive. It's getting hard to find a retread in many areas today because the price advantage is marginal.

Which tire for you? A radial if you are willing to put up with a harsher ride to get better handling and long tire life. Radials cost more initially, but are cheapest in the long run.

A bias-ply tire is your best bet if you don't intend to operate your car long enough to get full use out of a radial. Bias-ply tires are less subject to sidewall damage and give a softer ride.

Belted-bias tires, available as original equipment on many cars, are a good compromise. They cost less than good radials, are almost as inexpensive to operate in the long run, and represent a happy median between ride and handling.

Tires with a 6-, 8-, or even 10-ply rating have a greater load-carrying capacity than a 4-ply tire. Their disadvantage is that they give a stiffer ride and build up more heat in high-speed operation. Unless you are hauling heavy loads, stick to a 4-ply rated tire and get a belted-bias or radial for high-speed turnpike cruising. Go to a 70-series tire from a standard 78 for improved traction and load-carrying ability at the expense of somewhat harder steering. Avoid ultrawide tires, weird tread patterns, and rear tires larger than front ones unless you have a special problem and have sought the advice of an expert in solving it.

Ready to buy tires without being cowed by know-it-all salesmen? Just a few more things to keep in mind. To the advertised price of a tire you must add not only a few more dollars for sales tax, but also another few dollars for a special federal tax on rubber. Some tire dealers charge extra for mounting the tires on your wheels while others don't (so ask first). Then there's the charge of from $2 to $4 which many dealers make for balancing each tire. This charge may be automatic unless you specify that you don't want the tires balanced. (For a consideration of where and how to have your tires balanced, read the balance and alignment section in this chapter.)

SNOW TIRES AND CHAINS

Snow tires should be put on before the first snowfall is anticipated and taken off only when the last one has almost surely passed. November to late April is the usual season in most North Atlantic states.

Keeping both snow tires mounted on wheels will save having to stop at a garage when you want to change. One snow tire can be used as a spare in emergencies although it is not a good idea to travel on it for long periods since the wheels may get out of alignment. You can often purchase extra wheels for your car at a junkyard for a few dollars apiece.

Snow tires are helpful on loosely packed snow but have little or no effect on hard-packed snow or ice. For these conditions studs are most effective. Most snow tires are manufactured with holes in which a garage with the proper equipment can install metal studs for $6 to $8 per tire. Studs have been frowned upon lately by the authorities in many states since they tend to tear up the road surface. However they are a good safety measure.

Studs must be inserted on a tire before it is used. Tires with even a few weeks' wear may not hold studs properly. Also, studded tires which are worn cannot be used as replacements for regular tires in the way that normal snow tires can.

Some tires without a snow tread can be studded. These are for the front wheels of conventional rear-wheel-drive cars or the rear tires of front-wheel-drive automobiles. Studs on all four wheels provide maximum braking power and control.

When studded tires are removed from their wheels, always mark the direction of rotation. Studded tires replaced on a wheel so that the direction of rotation is opposite from what it previously was will exhibit uneven stud wear and possibly lose some studs.

Chains are even more effective than studs on ice and hard snow. A sturdy set of reinforced (sometimes called "V-bar") chains will provide the best traction and braking power under these conditions. The trouble with chains is that they are a nuisance to put on and take off, and they break easily when driven on dry roads or used extensively at speeds of forty miles per hour or more.

There is a fastening device which comes with most sets of chains and supposedly allows you to put them on without jacking up the car. You use the device to hold the chain to the wheel, then drive forward a little to wrap the chain around the wheel. My experience is that these fasteners do not work too well and may either fall off or allow the chain to wrap itself around the axle.

A good method of putting chains on is to jack the car up with the hand-brake on to keep the wheels from rotating. Then spread the chain over the wheel and fasten it on the inside first and then the outside. It is necessary to pull the chain taut for a tight fit and this may take some strength. If the chain is not tight enough when the catches are fastened to the end links, use the next link. Rubber "octopus" holders hook on the chain at five points or so on the outside of the wheel and should always be used to keep the chains from slipping off. Octopus holders usually do not come with a set of chains and must be purchased separately.

When chains break they can be repaired with spare links which either screw shut or fasten themselves when you run the car over them, causing the ends to be pressed tight. This is a temporary repair since the chain, once broken, is likely to break in another place.

Chains can be taken off without jacking by stopping so the chain latches are not under the wheel. Loosen the inner latch first, then the outer one. Lay the chain flat and drive off it. Wash and dry the chains before putting them back in their box since road salt is highly corrosive.

There are so-called "monkey chains" which fasten to the wheel in the same way crampons fit on shoes. Most types require a slotted wheel to fasten to and the hub cap may have to come off. Monkey chains are easier to put on but they are not very effective and tend to loosen and slip. This type of chain is best used to gain traction for a short time—such as to get up an icy hill—rather than for extended use.

TIRE REPAIRS

In routine servicing it is unlikely that you would want to go to all the trouble of breaking a tire bead, levering the tire off its wheel rim, patching it, and expanding it in place on the rim again—all to save the $2 or so any gas station would charge to do the job. Therefore the step-by-step repair of a flat tire is covered as an emergency procedure in Chapter 8.

Tubeless tires can be repaired with a rubber plug inserted from the outside with a needle-like tool or plug gun, without bothering to remove the wheel from the rim. This repair is only temporary and dangerous for high-speed driving.

The big problem in repairing modern tubeless tires is that they are sealed to the rim with a steel band (known as a "bead" since it is shaped somewhat on the style of a beaded necklace) which is mighty tenacious. To break the grip of the bead requires a special tool which works like a giant nutcracker, or a bead breaker working on the battering ram principle and constructed as a large, chisel-nosed slide hammer.

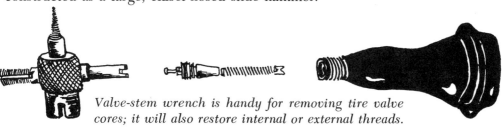

Valve-stem wrench is handy for removing tire valve cores; it will also restore internal or external threads.

Tire bead breaker. (Ford)

Tire bead expander. (Ford)

In addition to a bead breaker, a bead expander is required to constrict the tire as it is being inflated, thus ensuring that it seats properly on the rim. Bead expanders are either metal bands with a ratcheting mechanism or rubber tubes which constrict by inflation.

Gas stations do the whole job in a jiffy with a pneumatic bead breaker/expander on a permanent mount. Unless you are so far out in the boondocks that you can't even send smoke signals to the A.A.A., fixing a flat is one job which is better left to your corner gas station.

REPLACING SHOCK ABSORBERS

To replace shock absorbers you unbolt the old ones and bolt the new ones on.

It's not as easy as it sounds because the shocks are under your car where they've been getting a liberal dose of road salts and other corroding substances. Chances are that the bolts are rusted on tight.

Beginning at the rear, inspect the shocks to see where they're attached. The tops may be attached through recesses in the forward part of the trunk compartment. To get at them, you will usually have to pry up little rubber plugs.

Aside from corrosion, one problem in loosening the shock bolts is that the shock piston will often rotate freely, giving you nothing to turn against.

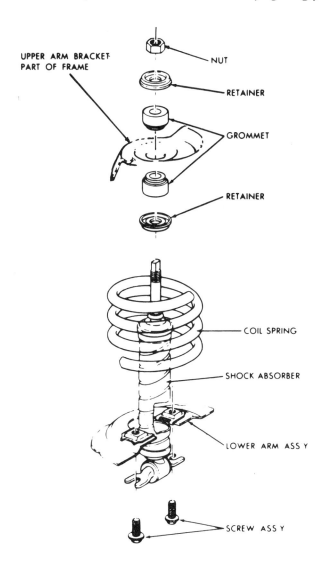

Exploded view of front shock mounting inside coil spring. (Pontiac)

The solution is to hold the tip of the piston shaft tight with (a) a special shock absorber tool or (b) a small socket which will fit or (c) a pair of vise-grips. If you don't care about saving the shock, an effective method is to hold the shock absorber body itself with the biggest pipe wrench you can find. Having an assistant to perform this chore while you loosen bolts up topside will be a big help.

Stubborn nuts can be attacked with the usual weapons. First penetrating oil, then heat (caution: stay away from the gas tank).

Next you can try chiseling the nut off, drilling it with your electric drill, or using a special tool known as a nut buster.

Special tools can aid in holding shock body against twisting while nuts are removed. (Owatonna)

Off with the old and on with the new. Follow the instructions with the new set of shocks for installing the replacement hardware. Lubricate the rubber bushings, washers, and nuts with silicone spray.

At the front you will generally find the tops of the shocks mounted with shock towers under the hood, and the shocks themselves located inside the front coil springs. Unbolt the shock towers before you jack up the car to work underneath.

Whenever shocks are placing tension on coil springs, be sure to place your jack stands so that the springs are compressed. At the front, put the jackstands under the lower control arms or jack at the frame when the upper arm is carrying the load. At the rear, place your support under the rear axle. This will prevent any accident which could occur when tension is taken off the springs.

The standard shocks supplied on most cars are anemic. When replacing shocks you will get better service from heavy-duty models, or the more expensive adjustable variety, than from original equipment replacements.

WHEEL BALANCE AND ALIGNMENT

There are two schools of thought on wheel balance and alignment. One says that you should find a good front end shop and leave it all to them. The other claims that wheel balance and alignment are well within the scope of the average backyard mechanic.

My view is that you can do it yourself provided that you are the finicky type willing to take the time to do each step correctly with no fudging. Professional help is necessary only when you run into problems beyond your scope. Your investment of $50 or so in special tools will be more than made up in the long run.

In the event that you do farm the work out, there are a few finer points which will enable you to get the most for your money. We'll run through these first, then go the do-it-yourself route.

Wheel balance will cost you $2 to $4 per tire no matter what kind of equipment is used to do the job. So, for the same price, you might as well go deluxe and patronize a place where a dynamic wheel balancer is used.

The process of balancing a wheel consists of attaching little lead weights to the rim to compensate for heavy spots in the wheel or tire. The simplest balancing device is a bubble balancer which works just the way it sounds. The wheel with its tire mounted is attached to the balancer and weights are attached to the rim until the bubble is centered.

The trouble is that this method does not accurately compensate for dynamic imbalance caused by centrifugal forces acting in different planes. A heavy spot near the outside of a rim travels faster than a heavy spot nearer the center and thus its centrifugal force is greater.

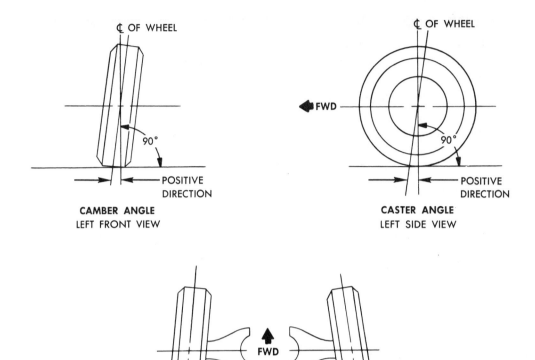

*Caster, camber, and toe-in, the three adjust-
ments for correct wheel alignment.* (Chevrolet)

Static balancing is about 75 per cent accurate and good for the retreads
on the old jalopy you use to putter around town in. On a fast car with wide
tires which is customarily driven at thruway speeds, dynamic balancing
pays off.

All machines which balance the tire while it is spinning are not true dynamic balancers. The device which balances a spinning wheel mounted on

the car is simply a more sophisticated version of a static balancer.

Wheel alignment is something of a black art complete with its own mumbo jumbo and awesome-looking machinery. Actually some pretty primitive equipment can be quite effective—it all depends on the operator. The man who does the job most effectively will check out the steering linkage, ball joints, front suspension, wheel balance, shocks, and tires before he makes a move to check the alignment. Then he will adjust everything to manufacturer's specs (although some really experienced operators have their own ideas and will adjust for apparent suspension problems and intended use).

When your car steers like a fat hog wallowing through the mud and the tires are scuffed or cupped, the ministrations of a real front end man are worth every penny they cost.

There are a few front end specialists in every city who are booked up for weeks in advance and do a lot of their work on cars brought in by other mechanics. Find the one nearest to you and you're all set.

Now for the do-it-yourself approach.

WHEEL BALANCE

Since dynamic balancing is beyond your scope, you'll have to stick to the old bubble balancer from Sears or J. C. Whitney for $20 or so.

Buy an assortment of wheel weights at the same time as you get the balancer. On many custom wheels the rim is too thick for conventional weights so you will need the stick-on variety with a roll of wheel weight tape. A special wheel weight tool is just a frill since a light hammer and pry tool will do the job as well.

Most bubble balancers come with detailed instructions. The first step will be to align some marks on the balancer. Then weights are placed on the rim and moved around until the bubble is centered. There's no formula for doing this, you simply swap around weights until things work out. It is easier to achieve balance with two smaller weights than with one large one.

When the bubble is centered, the weights are slid apart a few more inches and their position (the center of each) is marked with chalk lines across the tire sides and the tread. The chalk marks are used as guidelines to install weights on the other side of the rim. The idea is to halve each weight and put half the weight on the outside of the rim, the other half exactly opposite on the inside.

With your own bubble balancer, you can afford to balance the wheels as often as you like. Dynamic balancing will be necessary only if wheel shimmy develops and cannot be corrected by adjusting tire pressures to specifications and checking front end alignment including wheel and tire run-out.

If you are paying to have the job done, don't let the tire shop talk you into having the wheels balanced whenever you buy a new set of treads. Instead, wait until the new tires are run in for 300 miles or so, giving them time to "set" into their balance pattern. Then have the balancing done at a front end shop where an experienced man will be running his practiced eye over your linkage and suspension.

ALIGNMENT

Before checking alignment there are some preliminaries which are absolutely essential if you want to come up with any valid results.

Check tire inflation and adjust the pressure to your norm. Inspect the tires and don't bother to go any further if one tire is seriously worn, bulged, or cracked. Instead, invest your money in new tires.

Jack up the car and spin the front wheels to be sure that the bearings are still in good shape. Then check the ball joints and other suspension parts for excessive wear. For more on this topic you may want to peek ahead at the section in Chapter 11 on checking out a used car.

Test the shock absorbers and inspect them for leaks. Remove any excess baggage from your car which is not normally carried. Eyeball the springs to be sure they are not noticeably sagging.

If your car seems to be listing to one side or the other, and it is fitted with torsion bars, check your shop manual for the correct adjusting procedure before going any further. It really does not pay to do the job at all unless you're going to be painstaking.

Another preliminary task is to check for run-out. Wheel and tire run-out just means wobble from side to side. If you could hold your index finger

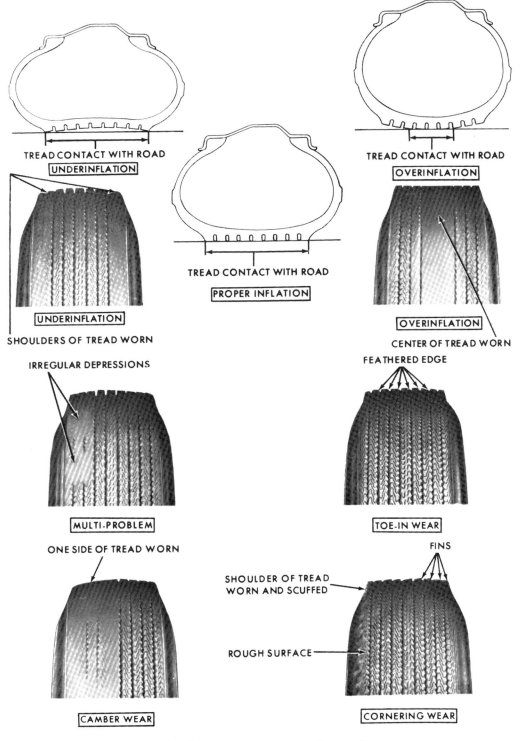

TREAD CONTACT WITH ROAD

UNDERINFLATION

UNDERINFLATION

SHOULDERS OF TREAD WORN

TREAD CONTACT WITH ROAD

PROPER INFLATION

TREAD CONTACT WITH ROAD

OVERINFLATION

OVERINFLATION

CENTER OF TREAD WORN

IRREGULAR DEPRESSIONS

MULTI-PROBLEM

FEATHERED EDGE

TOE-IN WEAR

ONE SIDE OF TREAD WORN

CAMBER WEAR

FINS

SHOULDER OF TREAD
WORN AND SCUFFED

ROUGH SURFACE

CORNERING WEAR

Chart for diagnosing tire wear. (Pontiac)

absolutely steady against a tire, and then spin the tire, any point where the tire was not in contact with your finger would indicate run-out.

Since your index finger is not really steady enough for the job, you'll have to find some suitable substitute. A nail driven through a board which is mounted on a sturdy base is the usual set-up and can be fabricated in a couple of minutes.

Spin the tires slowly as you check the wheels for run-out. There's no real cure for excessive run-out (wobble) other than replacing a warped wheel. A run-out of more than one-sixteenth of an inch means that you need new wheels.

When you have assured yourself that all component parts are shipshape, you're ready to do the alignment itself. You will be checking caster, camber, and toe-in.

Be sure your car is absolutely level as you go through these checks. Check the level of your working area with a carpenter's bubble level and use boards under the wheels to make any necessary height adjustments.

Toe-in is the difference of the distance between your front wheels at the front and the distance between them at the rear. In effect, your car is slightly pigeon-toed to correct for flexing of the metal as it moves forward. This toe-in is only a fraction of an inch.

Camber is the vertical tilt of the wheel, inward at the top or outward. Wheels that tilt outward at the top have positive camber and those that lean inward have negative camber. Most cars are set for a fraction of a degree of positive camber. This is because, on particular suspension systems, it keeps the tires tracking straighter and makes steering easier.

Caster used to be explained as the angle of kingpin inclination. Now that most cars have ball joints rather than kingpins, you can think of caster as the angle of inclination which an imaginary line drawn through the ball joints makes with the true vertical when viewed from the side of the car. Anyway, don't worry about it, since positive caster makes the wheels of rear-wheel-drive cars more stable while front-wheel-drive cars have negative caster for the same reason.

It's easy to make your own device for checking toe-in. You could just measure the distance between the leading edges of the front wheels with a tape measure, then take the same measurement for the trailing edges. Subtract the front measurement from the rear measurement and you have the toe-in, which can be checked against the specifications listed in your shop manual or a reference book such as *Motor's Repair Manual.*

Since the toe-in is only a fraction of an inch, doing an accurate job with a tape measure is plain frustrating. A much better idea is to use two thin, rigid boards, slide one along the other until the ends just touch the inner leading edges of your front tires. Make a mark on your boards showing this measurement. Then compare it with the same measurement taken at the trailing edges.

Of course, the wheels have to be pointing dead ahead when you take your measurements. Set your steering wheel straight and roll the car slowly ahead to be sure the wheels are straight before you begin your measurements.

If toe-in is not correct, it can be adjusted by loosening clamping bolts on the steering tie-rods and turning the rod sleeves to lengthen or shorten each tie-rod. To keep the steering centered, lengthen or shorten the tie-rod evenly on either side of the car.

There is a special tool for turning the rod sleeves, but you can do the job with a pipe wrench if you're careful. Don't use brute force on corroded threads. Try penetrating oil, heat from a propane torch, and prayer, in that order, to free up the sleeves.

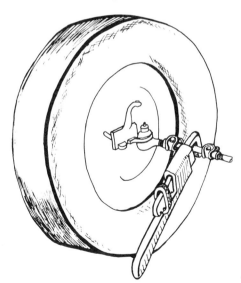

Adjusting toe-in. Special tools are preferred for use on tie rod sleeves, but pipe wrench will do if used carefully.

A commercial gauge for checking caster and camber costs about $15 and comes with detailed instructions. It is a bubble level held against the center spindle of each front wheel. For a camber check you take a direct reading with the wheel pointed straight ahead. The caster reading includes a procedure for turning the wheel a certain number of degrees outward before you take a reading, then turning it the same number of degrees inward for a second reading which is subtracted from the first. Although the tool can be clamped in place as you turn the steering wheel, the job is really easier with two people.

Again you will need your shop manual for caster and camber specifications. While you have the book out, check the adjustment procedure. Late-model Ford and GM cars use shims inserted at the sides of the upper suspension arm inner ends. Caster is changed by inserting shims at one pivot bolt only, while camber adjustments require the addition of an equal number of shims at either bolt. Chrysler and American Motors use eccentric adjusting bolts, while many models of VW (but not all) have had no

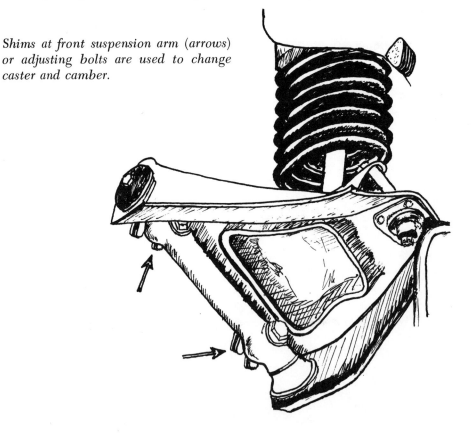

Shims at front suspension arm (arrows) or adjusting bolts are used to change caster and camber.

camber adjustments specified.

Better read that shop manual, and have your propane torch at hand to free stubborn bolts, and level off that garage floor. You can do your own wheel alignment for sure, but it's a job that'll have you sweating a little.

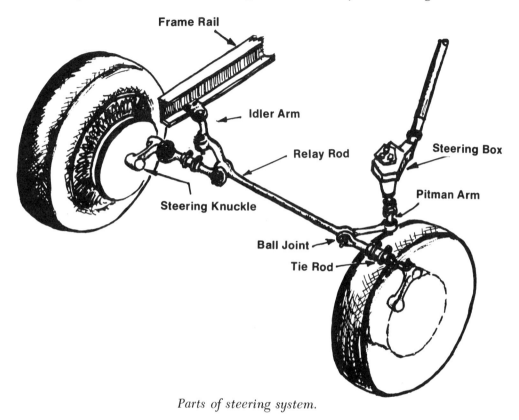

Parts of steering system.

MANUAL STEERING GEAR AND BALL JOINTS

The steering gear on your car will show signs of a coming rebellion by becoming very hard or excessively loose.

Your first line of defense against hard steering is to check the lube level. Most manual steering boxes have a filler plug which is also used to check lubricant level. Unscrew the plug and stick your finger in to see if the box is full up to the plug. On a few steering units there are cover bolts which serve as level check and filler plugs.

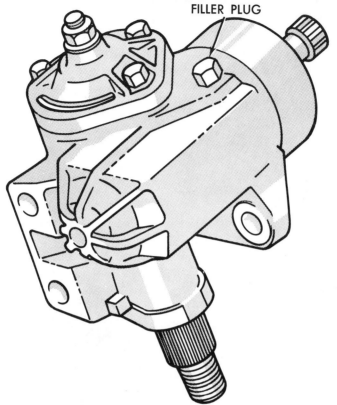

FILLER PLUG

Steering box has filler plug for adding lube. (Dodge)

Consult your owner's manual for the recommended lube. Hypoid gear oil of the SAE 90 variety is the usual stuff. Some boxes are filled with lighter oil. Substitute brands if you like, but always use the recommended type of lubricant.

An all-purpose oil suction gun is used to fill steering boxes as well as differentials and manual transmissions. Fill the box only to the plug level.

Loose steering, with a free play of more than an inch or so, can usually be cured by adjustments designed into the steering mechanism. On the recirculating ball and worm and roller varieties of steering—by far the most common types—there are two adjustments to be made. One controls the worm bearing preload, the other is the lash adjustment which controls the gear mesh depth.

Adjusting the steering gear yourself presents two problems. The Pitman Arm must be disconnected from the steering linkage, which requires a special puller, and the adjustments are made with a torque wrench reading in inch/lbs. Sure, many mechanics don't bother with a torque wrench and do it all by feel. Don't use this method unless you've had lots of experience.

It is also important to replace the Pitman Arm correctly and use new cotter pins where applicable. Since steering gear is so vital to driving safety, have the job done by a pro and you'll feel more secure.

Although steering adjustments can be left to the experts, you should learn to check ball joint wear yourself because ball joints are often replaced unnecessarily by dishonest or incompetent mechanics.

All too often a mechanic will put a car up on a lift, jiggle a front wheel up and down, and solemnly swear that the ball joints are worn and must be replaced immediately or the wheels are likely to fall off at any moment. The amount taken in all over the country by this little racket is probably enough to buy out Howard Hughes.

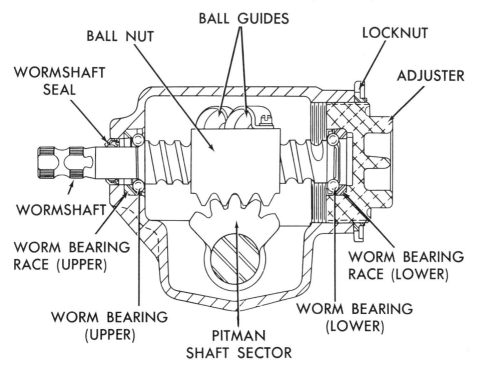

Cross-section of steering gear. This is the system known as worm and sector. (Chevrolet)

It is true that a worn ball joint could pop out of its socket and cause big trouble. The questionable part is the method of measuring wear and determining how much is dangerous.

The only standards are those set by the car manufacturer for the particular suspension design. Some specify that joint play be measured by checking axial (up and down) movement of the wheel assembly, while others require an in and out (tire sidewall) measurement. Allowable play varies from about five-hundredths of an inch, which is usual, to a full quarter-inch on some vehicles. You'll have to check a shop manual for the specifications on your car.

Look under the car at the front and you should see two heavy steel arms going to each wheel. At the end of both the upper and lower arm is a ball joint. Ball joints allow the wheels a necessary degree of movement just as the hip joints in your body keep you swinging loose.

To do the job correctly, you need a dial gauge which mounts on the wheel assembly. An inexpensive one ($15 or so) will be a good long-term investment.

Excessive ball joint wear usually becomes troublesome before the danger point is reached. The front end is thrown out of alignment, tire wear becomes irregular, and wheel shimmy may become apparent. At this point, if you don't want to do your own troubleshooting, bring the car to that super-good front end shop previously mentioned. Watch how they measure ball joint wear on your car and you'll never again be conned by the wheel jigglers.

Let the front end shop do your ball joint replacement also. Special tools are required, especially for those ball joints which are riveted on, and it's a hassle.

One ball joint at each wheel is the follower joint, while the other carries the load. The follower joint is kept preloaded so that the bearing surfaces are in constant contact. Follower joints can wear, but they wear much more slowly than load-carrying joints. Therefore the procedure for checking ball joint wear always refers to the load-carrying joint.

To determine which is the follower and which is the load-bearer on your car, look at the coil spring. If the spring is attached to the lower arm and runs between the two arms, then the lower arm is the main load-carrying member and play is measured at the lower ball joint.

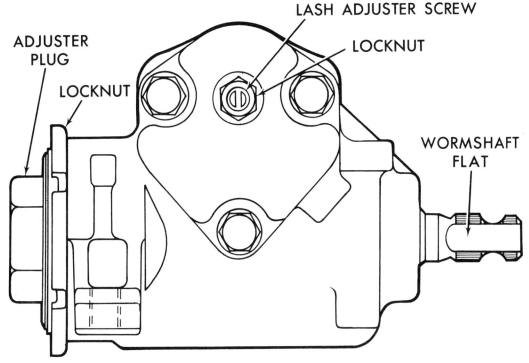

ADJUSTER PLUG

LOCKNUT

LASH ADJUSTER SCREW

LOCKNUT

WORMSHAFT FLAT

Typical adjustment points for steering gear. (Chevrolet)

Should the coil spring be located above the upper arm and attached to it, then the upper ball joint play is measured.

Perhaps there is no coil spring at all. In that case your car has torsion (twisting) bar front suspension and the lower arm is the weight-carrying member.

Play at the ball joint must be tested with the weight removed from the load-carrying arm. If the lower arm is carrying the load, then place your jack under it. Jack under the frame or use a bumper jack to remove weight from the upper arm.

Now consult your shop manual to determine the correct procedure for measuring play. If axial play is specified, use a lever under the tire to move the wheel assembly up and down. For checking tire sidewall movement, push the tire and wheel assembly in and out.

Unless you are gifted with an extraordinary type of sixth sense, you

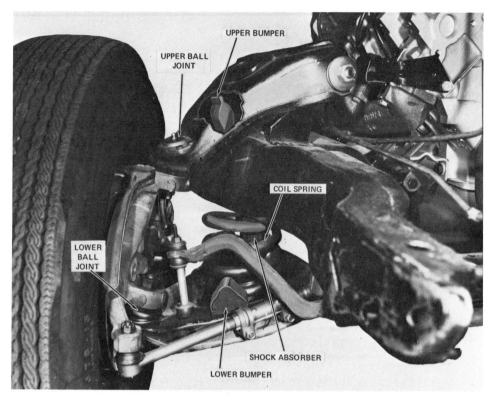

Front suspension components. (Pontiac)

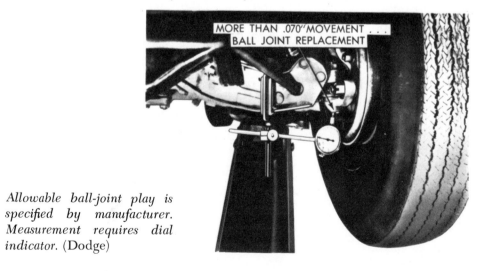

Allowable ball-joint play is specified by manufacturer. Measurement requires dial indicator. (Dodge)

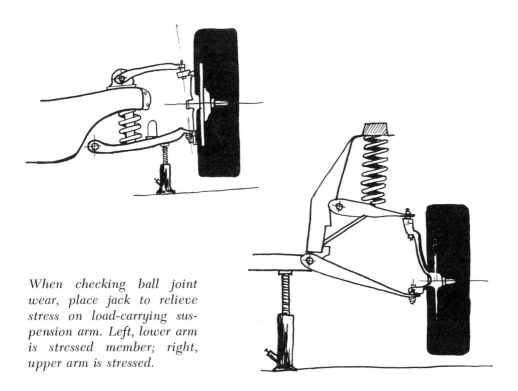

When checking ball joint wear, place jack to relieve stress on load-carrying suspension arm. Left, lower arm is stressed member; right, upper arm is stressed.

won't be able to determine whether play is a little less or more than five-hundredths of an inch. All you can say is that the ball joint feels pretty tight or kind of loose. So go buy that dial gauge if you want a definitive answer.

CHECKING BRAKE LININGS

The brake linings are of asbestos or metallic composition. They are about three-sixteenths of an inch thick and the material is long-wearing and heat resistant. Some linings are attached to the brake shoes by rivets while others are bonded on.

You won't get an argument from me on whether riveting is better than bonding, but some mechanics swear by riveted linings because there are quite a few cheap bonded linings around.

The disadvantage of riveted linings is that the rivets may be exposed to

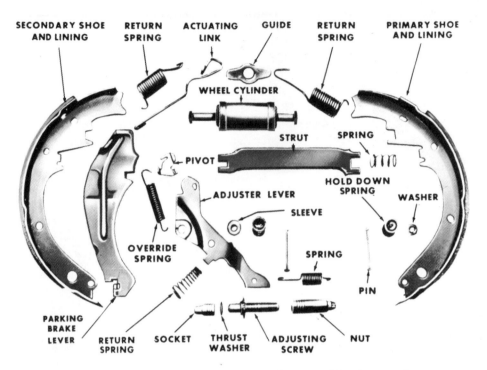

SECONDARY SHOE AND LINING · RETURN SPRING · ACTUATING LINK · GUIDE · RETURN SPRING · PRIMARY SHOE AND LINING

WHEEL CYLINDER

STRUT · SPRING

PIVOT

ADJUSTER LEVER · HOLD DOWN SPRING · WASHER

SLEEVE

OVERRIDE SPRING · SPRING

PIN

PARKING BRAKE LEVER · RETURN SPRING · SOCKET · THRUST WASHER · ADJUSTING SCREW · NUT

Components of self-adjusting brake assembly. (Pontiac)

the brake drum as the linings wear very thin. Drums scored deeply enough by exposed rivets can pass the point of being resurfaced (turned down smooth on a lathe made for the purpose) and have to be replaced. At $20 or so per drum this can get expensive.

Whether you have bonded or riveted linings, it is an essential safety precaution to inspect them once a year (every 12,000 miles or so) to be sure the linings are not worn to a thickness less than one-sixteenth of an inch at any point.

You do not have to inspect the brake linings at all wheels; one will do. Since front wheels must apply about 65 per cent of the total braking force, they can be expected to wear faster than those at the rear. The front brakes are applied equally and should wear at the same rate if the braking system is functioning well. Thus, inspection of the lining at either front wheel gives a good indication that all brake linings are holding up.

To view the brakes the whole wheel must come off at the wheel bearings. First remove the outer hubcap. Under the hub is the wheel-bearing grease cover at the center of the wheel. There is a special plier-like tool for gripping this component and pulling it loose, but you should be able to get it off easily by prying at either side with two screwdrivers.

Under the grease cover is the wheel-bearing assembly held in place by one or two large nuts. The outside is held tight by a cotter pin. Squeeze the ends of the cotter pin closed with pliers and pull it out of its slot. Loosen the nut or nuts with a large combination wrench or with an adjustable wrench.

Now jack the car up and place a jack stand or other firm support under it for safety. Then grab the wheel firmly with both hands and alternately tug and push at it until the outer wheel-bearing assembly pops out. Further tugging will remove the wheel and brake drum together.

A stubborn wheel will require that you poke a screwdriver through the brake backing plate on the inside of the wheel (or outside of the wheel on some cars) and move the brake adjusting wheel until the drum is freed. (We will get into the procedure for adjusting brakes shortly.)

With the wheel and brake drum off, you can inspect the brake lining. Also run your hand over the inside of the drum to feel for grooves. A grooved drum will need to be resurfaced. This job costs little and is best done by a professional mechanic with the proper equipment.

Wipe away the lining dust with a dry rag and replace the drum, wheel bearing, and fastening nuts. Always use a new cotter pin to retain the outer nut as cotter pins are weakened in removal.

Since taking the wheel off requires some effort, you may wish to repack the wheel bearing with grease at the same time as you inspect the lining. Repack the other front wheel bearing at the same time.

FRONT WHEEL BEARINGS

The rear wheel bearings in a conventional rear-wheel-drive car are sealed for life and do not require greasing as a routine maintenance procedure. Front wheel bearings are quite different. They need to be repacked with grease about every two years or 24,000 miles. When the job is not done and the bearings start to run dry, the front wheels develop more rolling resistance. Eventually the wheels may seize up.

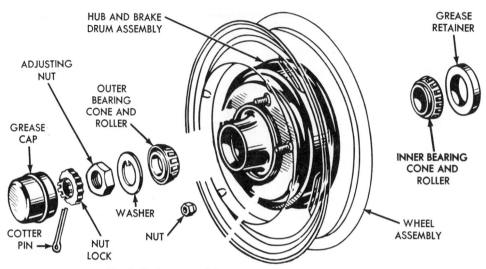

HUB AND BRAKE
DRUM ASSEMBLY

GREASE
RETAINER

ADJUSTING
NUT

OUTER
BEARING
CONE AND
ROLLER

GREASE
CAP

INNER BEARING
CONE AND
ROLLER

WASHER

COTTER
PIN

NUT
LOCK

NUT

WHEEL
ASSEMBLY

Exploded view of front-wheel bearings. (Ford)

You can check the condition of the front wheel bearings by spinning the tire and wheel with the car jacked up. A popping or grinding noise indicates that the bearings lack grease and may be worn or out of adjustment.

Even if the wheel spins freely without noise, the wheel bearing should be repacked every 24,000 miles as a matter of routine maintenance.

When the wheel came off, the washer and outer bearing came with it. The inner bearing is a press fit in the hub. Lay the wheel flat with its outer side facing you. Use a soft brass punch or a piece of wood and a hammer to drive the inner bearing from the hub. Place the punch on one side of the lip of the bearing and tap it gently with the punch until the bearing moves a little. Then place the punch on the other side of the lip. In this way you will keep the bearing going as straight as possible while you drive it from the hub. The bearing and washer-like grease retainer will drop to the floor.

Inspect the bearings for roughness and pitting and replace them if necessary. If the bearings seem smooth, wash them in kerosene to get all the grease out and dry thoroughly.

Wheel bearings use lithium or sodium grease readily available from auto parts stores. Use your hands to force grease through the bearing. Put a glob of the stuff in the palm of your hand and keep pressing the bearing into the grease until it is thoroughly packed. Pack all bearings.

Turn the wheel over so its inner side faces you. Place the inner bearing into the hub, narrow end first. Using your soft punch on both sides of the lip of the bearing alternately, tap the bearing down until it is almost flush with the hub. Dip the new grease seal in oil and put it over the end of the bearing (smooth side toward you) and continue tapping with the punch until the grease seal is flush with the hub.

With the older ball-bearing type of wheel bearing, the outer bearing is placed on the hub before the whole works is mounted on the wheel spindle. On the modern/roller-bearing type, the bearing and washer are not installed until after the wheel has been positioned on the spindle with the inner bearing in place.

Now the spindle nut (or nuts) is placed back on and tightened. This can be a little tricky. The nut needs to be tight but not too tight. And it must line up so the cotter pin will hold it to the spindle. The best way to do this is to have a torque wrench which measures tightening force and to tighten to specifications in the shop manual.

For those without a torque wrench, I suggest the following procedure. Use the box end of a large combination wrench (or even an adjustable wrench) to tighten the nut as much as you can using hand pressure alone. Then back off at least one full turn, or more if necessary to line up the cotter pin slot.

Always use a new cotter pin when you put a wheel back on, and bend the ends flat. Clip one end of the cotter pin short so that it will not be reused. Snap the grease cap back on and install the hub cap.

ADJUSTING BRAKES

The braking system includes a mechanism which compensates for lining wear by moving the brake shoe position slightly closer to the drum. On most newer cars this mechanism is self-adjusting and works whenever you apply the brakes while reversing. On a few cars it works when you apply the handbrake.

When the brake pedal grab point is gradually lowered over a period of time, suspect the adjusting mechanism. Back up a few times and apply the brakes vigorously, or pull up the hand brake if you are not accustomed to using it. If the pedal remains low, check the whole system for leaks as well as maladjustment.

Cars with manually adjusting brakes will require attention whenever the pedal is low. On most cars the adjustment is made by turning a knurled wheel known as the "star-wheel adjuster." A few older cars have a square lug nut on the inside of the wheel (brake backing plate) which turns with a wrench to move the brake shoes.

The star wheel can usually be reached by poking a screwdriver through a slot in the brake backing plate (you will have to remove a rubber plug from the slot first). Use the screwdriver to rotate the star wheel until you

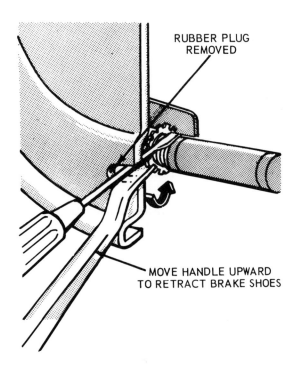

RUBBER PLUG
REMOVED

MOVE HANDLE UPWARD
TO RETRACT BRAKE SHOES

Adjusting brakes. (Ford)

feel the brakes dragging against the drum when you manually turn the tire. Get into the car and apply the brake pedal hard to center the shoes. Then try turning the tire again. If the brakes don't drag, rotate the star wheel a few more notches. Keep on until you can feel the brakes dragging when you turn the tire after stepping on the pedal. Then back off on the star wheel a few notches until the tire rotates freely.

Some star wheels can be reached through a slot in the front of the wheel (older VW's, for instance) while others have no slot at all so that the brake drum has to come off before you can adjust the shoes. Sometimes there are two star wheels; one to advance the drums and another to back them off.

When self-adjusting brakes do not work, you can manually adjust them using the same procedure as above. The only difference is that part of the self-adjusting mechanism may block the star wheel and have to be held to the side with another screwdriver as you rotate the star wheel with the first one.

Brake adjustments are made easier with a star wheel adjusting tool (sometimes called a "brake spoon") which has a bend at the end and a wide blade. Sometimes a brake spoon will get at a star wheel where a screwdriver will not and it is a very inexpensive tool.

BRAKE SYSTEM LEAKS

When your brakes begin to fail and the red warning light on your dashboard flashes on, look for a leak in the system.

Many people are under the impression that the brake warning bulb only lights up when they have inadvertently left the handbrake on after starting up the car. Actually it serves a dual function (on some cars there are two separate warning lights). When it flashes on and the handbrake is off, hydraulic pressure is dangerously low. There is a brake warning light on all late-model cars.

The characteristic sign of a hydraulic leak is a brake pedal that slowly sinks to the floor under steady pressure.

All cars now have a dual master cylinder. When one piston in the cylinder is not operating right, the other one will still give braking power at two wheels. Nevertheless, it's easy to determine that something's gone awry since the pedal goes down and the warning light comes on.

Check for external leaks first. Have someone apply the brakes to build up hydraulic pressure as you look around the master cylinder and have a squint at the brake lines running to the cylinder in each wheel.

Brake lines are made of steel or reinforced neoprene. They should never be replaced with copper tubing, which is not strong enough. The brake line connections are either pressure fittings or flared fittings. On the latter

Double flare is used for secure connection on pressure lines. (Chevrolet)

type, the steel brake line is flared out at the end and then in again, making a very strong, double-flared junction.

Replacing a leaky brake line is easy enough, but double-flaring a steel brake line with a flaring tool is a job that requires a definite knack. Unless you have mastered the technique, it is far easier to buy the brake line already made up with fittings for your car. It isn't always easy to find "ready-made" brake lines, but it's worthwhile searching a bit.

Never modify brake lines by using a different type of tubing or running it a different way. This is asking for trouble. Neoprene tubing can easily fail if it's run too close to a hot exhaust system, while steel line may crack if it is substituted for neoprene in an area where flexing is necessary.

A leak at a connection may sometimes be repaired by tightening the pressure nuts a bit with an open-end wrench. Some mechanics will attempt to stop the leak temporarily by applying a sealant such as Permatex at the junction. Applying sealant is an emergency measure only and it would be foolhardy to drive any farther than a nearby service station with the brake system in this shape.

Look for external leaks around the master cylinder and on the inner side of the wheels where the wheel cylinders are located. If the leak point is not apparent by simple eyeballing while someone puts pressure on the brake pedal, you'll have to check further.

Pull back the rubber boot (known as the dust cover) at the pedal rod extension end of the master cylinder piston. A few drops of brake fluid by-passing the cup and getting into the rubber boot is normal. A greater quantity means that you've found the leak.

In order to check each wheel cylinder, you have to pull the wheel off first. There's no easier way. The wheel cylinders are inspected in the same manner as a master cylinder. Pull back each rubber boot (at both ends on most cylinder styles) and look for a quantity of fluid by-passing the piston cup.

While you're checking the wheel cylinders for leakage, you might as well see if the pistons are moving freely in the cylinders without sticking. Detach the brake springs from the shoes (Vise-grips will do the job if you don't have a pair of brake spring pliers) and move the shoes away from the wheel cylinder. Then press each piston in to see that it depresses smoothly.

The only cure for a brake cylinder leak or a sticking piston is to rebuild the cylinder. A rebuild involves the replacement of internal parts—pistons, springs, cups, and washers—and is not very difficult. Rebuild kits for wheel cylinders cost only a few dollars, with master cylinder kits running a bit more.

There is one case where a defective master cylinder will exhibit all the symptoms of a leak, yet no leakage is visible. This is due to an internal leak at the master cylinder primary cup in front of the piston. The secondary

Exploded view of brake wheel cylinder. (Ford)

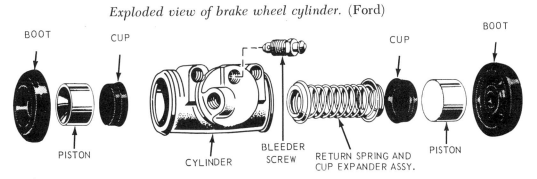

BOOT   CUP   CUP   BOOT

PISTON   BLEEDER   PISTON
               CYLINDER   SCREW   RETURN SPRING AND
                                    CUP EXPANDER ASSY.

cup still retains the brake fluid and prevents external leakage, but the fluid is not giving hydraulic pressure where it should. Again, the cure is to rebuild the cylinder, so let's move on to that task.

REBUILDING MASTER CYLINDERS AND WHEEL CYLINDERS

Whatever rebuild kit you buy should come with a diagram showing the order in which the parts are assembled. It certainly helps to simplify things since all you have to do is take the cylinder apart, replace old parts with new ones in the kit (always use them all), and follow the diagram when reassembling.

To remove a master cylinder, start by disconnecting the brake fluid line (two of them on a dual system) which runs to the cylinder. Don't worry about fluid leaking out since there's a check valve where the lines connect.

Next, pull off wires from the stoplight switch to the cylinder (if any) and then take your wrenches inside the car and unbolt the master cylinder pushrod from the brake pedal. Now all you have to do is unfasten the bolts holding the master cylinder in place and remove it.

At the front of the cylinder there's a snap-ring or metal collar held on with screws. If the retainer is a snap-ring, use a pair of snap-ring pliers to pull it off. In case you don't have these special pliers, the snap-ring can be pried off with caution. It is under tension and can spring up and crush the cigarettes in your shirt pocket, or whatever, when released. So hold something between you and the snap-ring as you pry at it.

Before pulling out the piston and cylinder internals, look for a retaining screw on the side of the cylinder body and remove it. Also, on master cylinders used with drum brake systems, there are check valves in the brake line inlets which must be removed. On a disc/drum system there will be only one valve, at the inlet of the line going to the drum portion of the system.

These check valves, also known as residual pressure valves, consist of a little valve and spring covered by a metal cap or tubing seat. The caps are removed by driving small, self-tapping screws into them and using the screws as a handle to pry against. The self-tapping screws are usually supplied with the rebuild kit in case you don't have any on hand. Tap the

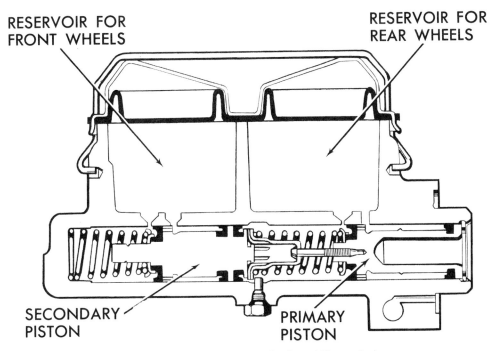

RESERVOIR FOR FRONT WHEELS

RESERVOIR FOR REAR WHEELS

SECONDARY PISTON

PRIMARY PISTON

Cross-section of master cylinder. (Chevrolet)

screw into the tubing seat gently with a hammer to get it started, then screw it down enough for a firm seat. Pry at either side with two screwdrivers and the tubing seat should pop out. You can then remove the little valve and spring with a pair of needlenose pliers.

The next step is to slide the internals out of the cylinder. On a dual system this is sometimes difficult since the rear piston may stick in the cylinder bore. If it refuses to come out by tapping the cylinder body or poking at the piston with a screwdriver, run down to your buddy at the service station and have him blow it out with compressed air.

Follow the diagram on the instruction sheet carefully when you reassemble the pistons with new parts. The only part of the job that is likely to present a problem is getting the pushrod loose from the front piston assembly if they are held together by a rubber bushing with a press fit. Try holding the pushrod in a vise and pounding on the piston assembly with a hammer and flat-nosed punch until it comes off.

Before you slide the rebuilt piston assembly back into the cylinder, check

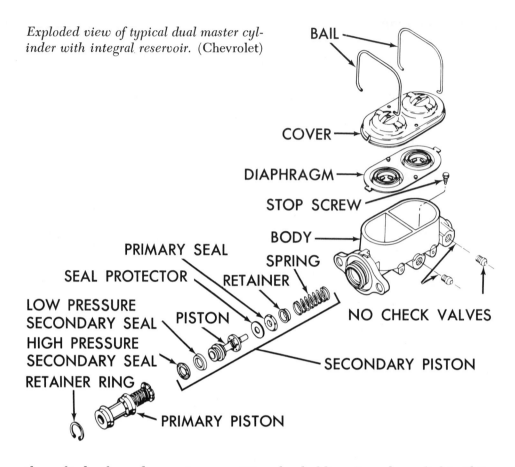

Exploded view of typical dual master cylinder with integral reservoir. (Chevrolet)

BAIL

COVER

DIAPHRAGM

STOP SCREW

BODY

SPRING

PRIMARY SEAL

SEAL PROTECTOR

RETAINER

LOW PRESSURE
SECONDARY SEAL

PISTON

NO CHECK VALVES

HIGH PRESSURE
SECONDARY SEAL

RETAINER RING

SECONDARY PISTON

PRIMARY PISTON

the cylinder bore for scoring or pitting by holding it under a light. If it looks pretty smooth, you can just clean it up a bit with a piece of crocus cloth (a very fine grade of abrasive paper) and then run a clean rag moistened with brake fluid through it. Never use solvent or oil to clean or lubricate any cylinder parts since lingering traces can contaminate the brake fluid.

A cylinder bore that looks rough must be honed. You can do this yourself with a cylinder hone which attaches to the end of an electric drill, but it's a better idea to have it done professionally. The maximum which can be removed from the bore is .005 of an inch of metal, and that's a pretty critical measurement. When the cylinder bore does need to be honed this much, you will have to make another trip back to the parts store and trade your rebuild kit in on one with a .005-inch oversize piston. An alternative is to buy a new master cylinder if the bore on the old one seems pretty rough.

Install the check valves and springs, then tap the tubing seats in place using a flat-nosed punch.

Now you are ready to bolt the reassembled cylinder back in, fill the reservoir with fluid, and bleed the system. Bleeding is covered in the next section.

The same general procedure is used to rebuild a wheel cylinder. Once the wheel is removed, detach the brake springs so you can get at the cylinder, then remove the rubber boot and pushrod (or boots and pushrods from a double-ended cylinder). Press against one of the pistons (or tap with a flat-ended punch) until the inner works comes out the other end.

On systems with a cylinder for each brake shoe (two per wheel) one end of the cylinder will be closed. Also there are some cylinders that are step-bored with a different diameter piston at either end. In this case the best way to remove a piston is by applying gentle pressure on the brake pedal to force it out hydraulically. Keep the pressure gentle or the pistons will come out like gangbusters and you'll have a mess.

Clean the wheel cylinder bores by running a piece of crocus cloth through them, then pull through a rag moistened with brake fluid to remove the residue.

Rough or pitted wheel cylinders can be honed with a device which chucks into an electric drill (it's a smaller hone than the one designed for master

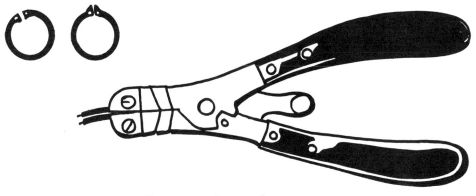

Snap-ring pliers and snap-rings.

cylinder bores), but again you should not remove more than .005 of an inch. Since wheel cylinders are inexpensive, it's usually easier to replace them when in doubt.

Reassemble components and lubricate with brake fluid before you fit the pistons in the cylinder. Connect the brake shoes and you are ready to bleed the system.

It's a wise idea to rebuild all the wheel cylinders when one of them fails due to old age. This is known as preventive medicine.

BLEEDING BRAKE LINES

Your foot hitting the brake pedal moves a piston in the master cylinder. The force is multiplied through leverage and applied to the fluid in the master cylinder, from which the fluid pressure is transmitted equally in all directions. As long as the brake lines hold the pressure without bursting, the force is channeled to each wheel cylinder and used to move the brake shoes against the drums.

If Archimedes lived today he might say: "Give me a brake pedal long enough and a system which will hold enough pressure, and I will stop the Earth."

When air gets in the hydraulic system, the brake pedal starts feeling spongy. This is because air compresses more easily than brake fluid. Whenever it happens, air must be bled off.

Air naturally gets in the system whenever the master cylinders or wheel cylinders are taken apart, so bleeding is always the last step of a cylinder rebuild.

There are brake bleed valves on the inner side of each wheel, near the wheel cylinder. Some master cylinders have a similar bleed valve. It is a small nipple similar to a grease fitting.

The procedure for bleeding is to crack the valve open with a small wrench (you may need an offset box-end wrench), after yelling to your assistant to tromp on the brake pedal and keep the pressure on hard, then close the valve and yell to him or her to let up the pressure. This process—open, tromp down, close, let up—continues until the fluid coming out the valve stops spitting and spluttering, indicating that there's no more air in it.

On master cylinders without a bleed valve, leave the fluid lines disconnected while your assistant tromps on the pedal and forces fluid out the check valve. Continue until the fluid flows without spluttering.

There are gadgets sold for bleeding master cylinders "on the bench." However this simple procedure will work with those having check valves.

Loner types without an assistant handy can purchase a length of tubing which fits snugly over the bleed valve, run the other end of the tubing into a jar of brake fluid, crack open the valve, and then work the brake pedal ten or twelve times. Having the tube submerged in brake fluid will prevent air getting into the system whenever the pedal is released. The

Bleeding brakes.

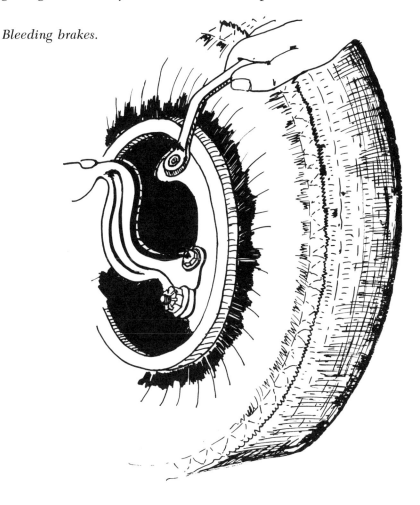

valve should be shut before the hose is pulled off. With four lengths of hose and four jars of fluid, you can even bleed all the brakes at once.

Professional mechanics sometimes use a pressure bleeder which does the whole job in one shot with little trouble, but this device is priced out of the amateur class.

Remember to always use brake fluid with a specification of 70-R3 or J1703A (it should say so on the label), since this is the only type with a high enough boiling point and other necessary characteristics to meet car manufacturer's specifications. Disc brake systems require a special fluid which is often dyed bright blue.

Prolonged exposure to air can ruin brake fluid. Never leave the cap off the can for a long time or re-use fluid that has been bled off.

Replenish the fluid in the master cylinder reservoir as you bleed the brakes. For convenience, leave the cover off the cylinder. The little geysers in the fluid reservoir which arise as pressure is put on the brake pedal show that the system is operating correctly. When you are done, fill the reservoir to about one-quarter inch below the brim and snug the cap down tightly.

(You did put the reservoir cover back on snugly, didn't you?)

Although it's not critical, in my experience, most manufacturers recommend that you bleed the wheel farthest from the master cylinder first, then the next farthest, etc.

Careful about getting brake fluid slopped all over. It does a nice job removing paint. If you do happen to get brake fluid on the brake linings, wash it off quickly with denatured alcohol.

Should you flush the brake lines and change fluid as part of routine maintenance? This seems to be a gray area. The people who make brake fluid would like you to change it often and boost their sales. Many people never change their brake fluid and nothing seems to happen. Perhaps this is because newer cars have an indirectly vented fluid reservoir so moisture can't get in and condense in the fluid.

Compromise and change all the fluid whenever you do a brake job or at least every third year. Unless the fluid shows signs of contamination (foreign particles, spongy brake pedal, rubber parts in the piston assemblies swelled up), you don't have to disassemble the cylinders and get every last drop of fluid out. Just bleed off and refill. If there is contamination, you will need to do a complete job with brake flushing fluid from your local auto supply store.

PROPORTIONING VALVES AND BRAKE LIGHT SWITCHES

On cars with front disc and rear drum brakes a proportioning valve is used to ensure that the drum system (which does not need as much pedal pressure to lock up the brakes) will receive only a portion of the pressure that the discs are getting. The valve goes into operation when a certain pressure is reached and then sends a part of the additional pressure to the rear brakes. The valve does not cut off all pressure beyond a certain limit. The pressure sent is always proportional to the amount your foot is putting on the brake pedal.

Despite the advent of proportioning valves—so handy that they are now used on some all-drum systems—brakes on one axle or another still do lock up prematurely at times. This is usually not the fault of the valve. A maladjustment in some other part of the system is throwing the proportions out of whack.

Troubleshoot the system by adjusting the brakes, having a mechanic check for out-of-round drums or brake linings which are not trued to the drums, and checking out the brake cylinders.

Where brake systems are extensively modified (oversized drum brakes or discs fitted in place of the standard drum brakes), the balance of the system may be lost. The least expensive way to balance brake forces and prevent lock-up is to adapt an adjustable proportioning valve. The one found on 1968 Corvettes with heavy-duty brakes is widely used by backyard Smokey Yunicks and can be ordered through any Chevy dealer who reads his parts book.

The two types of light switches associated with brake systems are the stoplight switch and the brake warning light switch which operates the red signal light on your dash. The stoplight switch is covered in Chapter 4.

Brake warning light switches are mounted on or near the master cylinder. They usually have two hydraulic leads and one electrical lead going to the cylinder. On GM and Chrysler cars the light switch is pretty straightforward, but on Fords there's a trick to getting the warning light off after you have repaired a leak or bled the brakes.

The technique is to open one bleed valve on the wheels, put steady pressure on the brake pedal, then quickly lift your foot from the pedal as the light goes out. Close the bleed valve and then top up the fluid in the master cylinder.

If the above method doesn't work, try again and this time open a bleed valve on the other side of the car. Which bleed valve to open depends upon the position of the piston in the master cylinder. It's simplest just to try both ways.

RELINING BRAKES

Relining brakes is a job which calls for close cooperation between you and a repair shop or machine shop which has a brake drum lathe. Brake drums often get scored from dirt or the bare rivet heads of worn lining, and they must be turned down on a drum lathe to be restored to a smooth and concentric condition.

The part of the job you can do is removal of the old linings and installation of new ones. Grinding down the brake drums and truing the new linings to them are chores that must be done professionally.

The first step is to put the car up on jack stands and take the wheels off. If the brake drums cannot be pulled off, back off the star wheel brake adjuster until the drums can be loosened. Sometimes the application of heat will help to remove a stuck drum. There are also special pullers for rear wheels.

Before proceeding further, bring the drums to the repair shop you have decided to use. Here the drums will be measured with a brake drum gauge to determine whether they can be turned down successfully. If the drums are too far gone, you may have to purchase new ones or hunt up acceptable substitutes in a junkyard.

Assuming that the drums can be cleaned up without removing too much metal, you're in business. Now all you need is a new set of linings.

Brake linings are usually purchased already riveted or bonded to the brake shoes. Most likely the shop which ground down the drums can supply the linings, either standard or oversize if necessary. Then they can grind the linings on a special machine to true them to the drum contours. Pick up a set of retracting springs to go along with your new linings and head home.

Now all you have to do is remove the old linings and install the new ones. Examine the brake assembly carefully before you take the old linings off so that you know where each component fits.

Since residual hydraulic pressure on the brake line could force the wheel cylinder pistons out when the shoes are removed, a wheel cylinder clamp should be fitted on each cylinder before going further. Newer cylinders often have a wheel cylinder stop plate, in which case clamps are not necessary.

Brake springs can be removed with a pair of vise-grips, but spending a few dollars for brake spring pliers will make things a lot simpler. The gadget on the handle of the pliers can be used to twist off some retracting springs, while the business end will release other types of springs plus the self-adjusting assembly spring at the bottom of the shoes. On certain de-

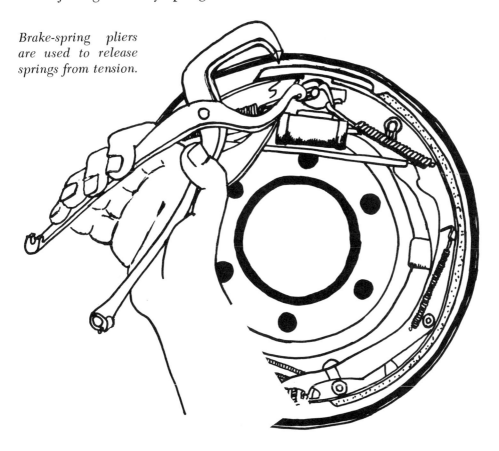

Brake-spring pliers are used to release springs from tension.

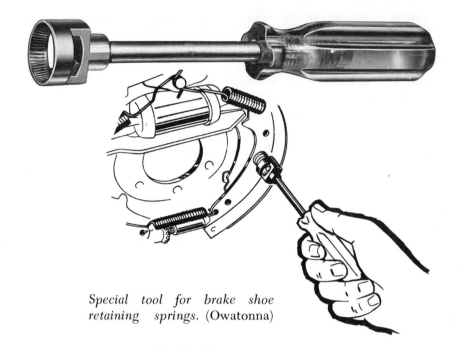

Special tool for brake shoe retaining springs. (Owatonna)

signs, both shoes and the adjuster assembly can be removed as a unit and they will then separate.

There is another tool used to unfasten the hold-down springs by pushing them in and giving a half-turn. Sometimes a thin screwdriver blade can be substituted.

While you are installing the new shoes, be sure not to get any grease on them or even touch them with your sweaty fingers. If you do happen to get grease on the linings, wash them off carefully with alcohol.

Clean the brake backing plate with solvent and clean and lubricate the adjusting assembly. Lubriplate or another high-temperature grease must be used. The brake shoe support pads should also be wiped with a clean rag and then lubricated with a thin coat of the same grease.

One problem you may encounter in installing the shoes is determining which is the primary shoe and which is the secondary one. On the single-anchor type of brakes the primary shoe will generally be shorter than the secondary. Some designs use interchangeable shoes and others employ shoes of the same dimension but different lining composition. If there is any doubt about which shoe goes where, ask at the repair shop when you buy the new shoes.

Put the shoes in place, install the hold-down spring on the secondary shoe, assemble the self-adjuster mechanism and be sure it lines up with the adjusting port in the wheel, and attach the primary shoe. On rear brakes, be sure to reconnect the parking brake cable and strut. Then hook up the retracting springs on the shoes and attach them to the anchoring pins.

Check the brakes to be sure the shoes move freely. Then wipe down the drums carefully to remove any metal particles and re-install them. Adjust the brakes until the pedal feels right.

It is not always necessary to turn down the brake drums, true the new linings, and install new retracting springs. Watch a professional mechanic and you are likely to see some shortcuts being taken. However, one of the reasons for doing the job yourself is to see that it's done right, and you'll be saving a substantial labor cost.

DISC BRAKES

There are disc brakes on the front wheels of quite a few American cars; some foreign cars have discs on all four wheels.

The disc, or rotor, is a circular steel plate which is attached to the wheel and rotates with it. On the disc is a caliper, fitting over one segment of the disc like a C-clamp. Hydraulically operated pistons push brake pads against the disc when the driver steps on the brake pedal.

The main advantage of the disc is its freedom from the effects of heat. Drum brakes "fade" when heat builds up through repeated application. The drum and brake shoes, buried deep within the wheel, heat up rapidly. With heat build-up the drums get out of round and the linings glaze and lose their bite.

Discs are exposed to the air and run cooler. But even when discs do heat, they don't fade. Disc pads have been developed to operate satisfactorily at high temperatures and contact with the rotor remains as effective. The factor which has prevented wholesale adoption of discs on all four wheels is that they are more expensive to manufacture and not as easy to fit with a handbrake mechanism.

Bleeding disc brakes is done in the same way as bleeding drums with

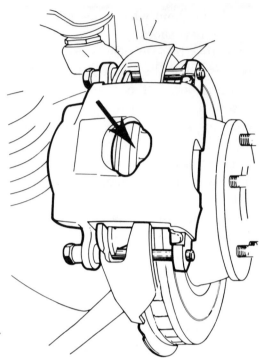

Disc brake assembly. Arrow shows lining inspection port. (Pontiac)

one exception. Many front disc brakes used in systems with rear drums have a metering valve somewhere in the brake line to the front calipers. The valve must be opened while bleeding the brakes and may require a clip to hold it open. The clip should be removed when bleeding is done.

The disc caliper mechanism is rather complex and requires expert attention when it fails to operate correctly.

All calipers on current production cars have a see-through part which allows a visual check of the disc pads without removing the caliper itself. Just take off the wheel, look through the inspection hole, and see if the disc pad has at least one sixteenth of an inch of lining left.

Replacing disc pads is fairly simple because it is only a matter of taking out the old disc pad and attaching the new one. Unlike drum linings the disc pads do not have to be trued to be sure they make contact at all points. The rotors rarely need resurfacing and no brake adjustment is required on most disc systems.

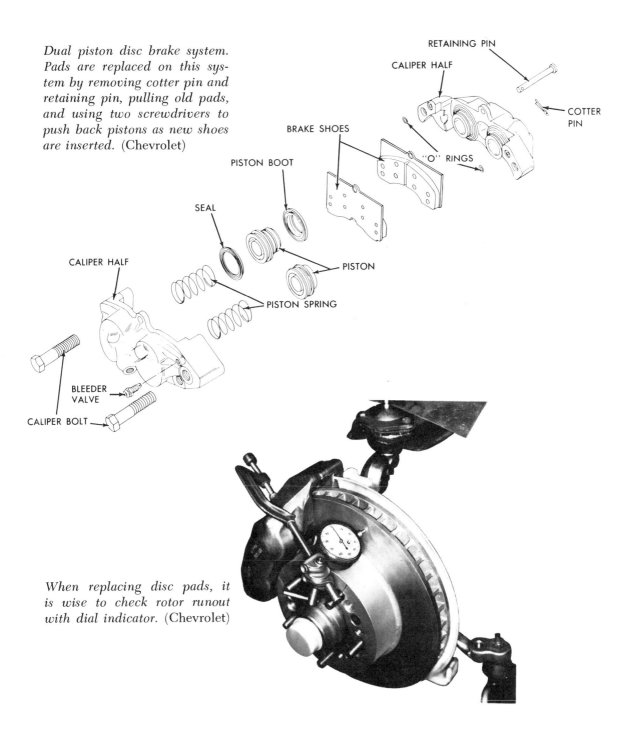

Dual piston disc brake system. Pads are replaced on this system by removing cotter pin and retaining pin, pulling old pads, and using two screwdrivers to push back pistons as new shoes are inserted. (Chevrolet)

RETAINING PIN

CALIPER HALF

COTTER PIN

BRAKE SHOES

"O" RINGS

PISTON BOOT

SEAL

PISTON

CALIPER HALF

PISTON SPRING

BLEEDER VALVE

CALIPER BOLT

When replacing disc pads, it is wise to check rotor runout with dial indicator. (Chevrolet)

Since the actual procedure for replacing discs differs from brand to brand, it is best to consult the service manual for your car before attempting the job.

POWER STEERING AND POWER BRAKES

Power steering is a hydraulically operated system to take the effort out of turning the steering wheel. It works by a belt-driven pump which supplies hydraulic pressure to the steering unit.

You can help keep the power steering mechanism in good shape by checking the fluid level, making sure the drive belt is in good condition and adjusted right, and watching for leaks.

When the fluid level goes down, the power steering pump can suck in air. You can bleed the system by simply turning the steering wheel from lock to lock rapidly a few times. Do this with the engine idling and some newspapers under the front wheels to make turning easier.

Power steering pump mounting. (Chevrolet)

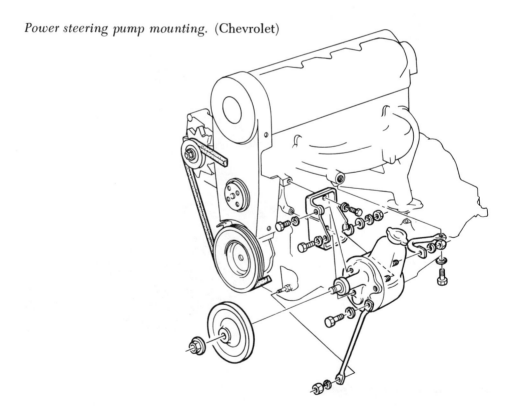

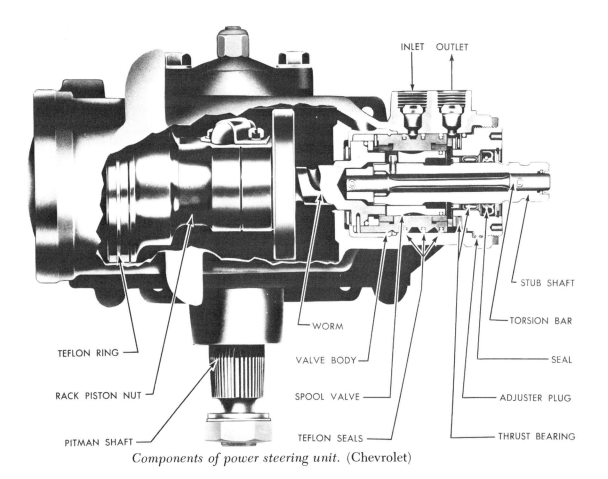

INLET   OUTLET

STUB SHAFT

TORSION BAR

SEAL

ADJUSTER PLUG

THRUST BEARING

TEFLON RING

RACK PISTON NUT

PITMAN SHAFT

WORM

VALVE BODY

SPOOL VALVE

TEFLON SEALS

Components of power steering unit. (Chevrolet)

Maintain belt tension at a point where the belt deflects about one-half inch at mid-point under finger pressure. Tighten the belt, when necessary, by loosening the pump mounting bolts and prying the pump to the position where belt tension is correct. Use caution when prying to avoid damaging the pump.

If your car has a tendency to make an aww-aww sound when the steering wheel is cut all the way to the left or right, it is probably a sticking pressure relief valve in the power steering system. Addition of a lubricant made for this purpose, such as Wynn's Concentrate, can get rid of the noise.

Check for leaks after you have cleaned grease and dirt off the power steering mechanism with engine degreaser. Turn the wheel from lock to

201 · *Chassis, Power Train, Body*

lock a few times and hold it against each lock for a minute. Then inspect for leaks. You can try an auto transmission sealing compound to get rid of a leak. If this does not work, take it to a garage. A leaky hose is easy to replace but a leaky shaft seal or power cylinder will require expert attention.

Power brakes are operated by a vacuum booster. The vacuum is drawn from the intake manifold to a power booster. There may also be a vacuum reservoir between the manifold and the booster. The booster works in conjunction with the brake master cylinder.

Some power boosters have an air filter which can be removed for cleaning. Otherwise the only maintenance required is to check the vacuum hoses to be sure they are tight and in good condition. Replace hoses that feel soft and spongy or excessively brittle.

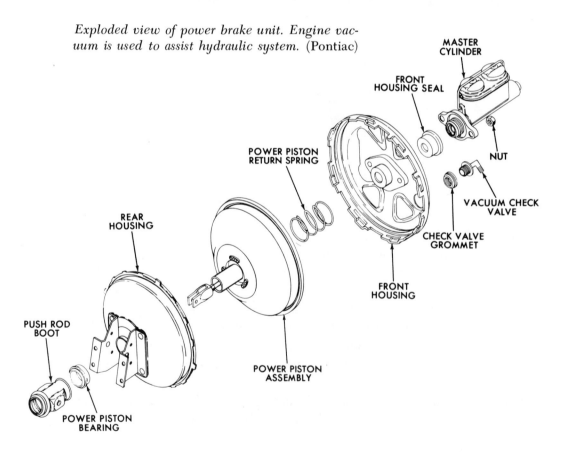

Exploded view of power brake unit. Engine vacuum is used to assist hydraulic system. (Pontiac)

MASTER CYLINDER

FRONT HOUSING SEAL

POWER PISTON RETURN SPRING

NUT

VACUUM CHECK VALVE

REAR HOUSING

CHECK VALVE GROMMET

FRONT HOUSING

PUSH ROD BOOT

POWER PISTON ASSEMBLY

POWER PISTON BEARING

You can check for a leak in the system by having someone start the engine and hold the brake pedal down while you listen for escaping air at the manifold, vacuum lines and the booster unit. A leaky booster unit will usually have to be replaced.

Other than the above, the power braking system requires no routine maintenance. Just keep the brakes themselves in good shape and keep the engine in tune so that the manifold puts out its full vacuum pressure. Check with a vacuum gauge if in doubt.

HANDBRAKE ADJUSTMENT

The conventional handbrake linkage consists of a cable running beneath the car and branching off to each rear wheel. By pulling on the lever, or

Adjusting handbrake in rear-wheel system (top) and internal drive system (bottom).

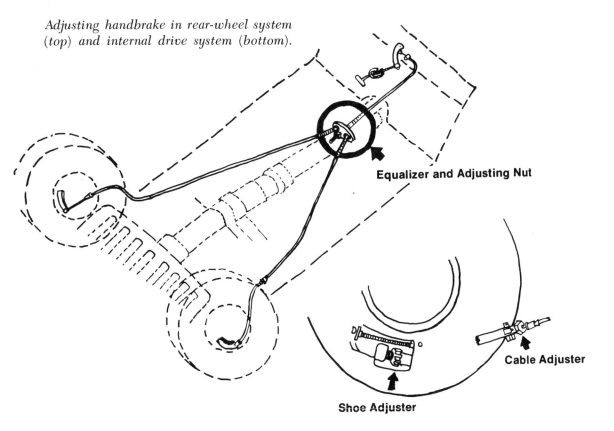

Equalizer and Adjusting Nut

Cable Adjuster

Shoe Adjuster

"umbrella handle," the rear brakes are mechanically applied and set by a ratchet mechanism.

The handbrake cable lengthens in normal use until it gets to the point where the brake no longer holds effectively. From time to time it is necessary to shorten the cable by taking up some slack with an adjusting nut.

On older cars without self-adjusting brakes, it is necessary to adjust the brake shoes on the rear wheels, by the star-wheel method discussed previously, before the handbrake linkage can be set correctly.

Most manufacturers specify that the handbrake lever be pulled up three or four notches before an adjustment is made. After doing this, get out your jackstands and wheel chocks so that you can work under the car safely.

Trace the handbrake cable beneath the car and you will find an equalizer mechanism at the "Y" where the cables branch off. On or near the equalizer is an adjusting nut.

Tighten the nut until drag on the brakes is felt when you spin a rear tire. Then slack off on the nut until the tire rotates freely. Finally, slack off a few more turns so that the cable has about two inches of slack in it. Test the handbrake to see that it holds well.

There may be a lock nut as well as an adjusting nut or screw on the cable mechanism in your car. In this case the procedure is to loosen the lock nut, make your adjustments, then tighten it again.

Chrysler Corporation is the maverick here. Older Mopar products were fitted with an external parking brake on the driveshaft, while the latter-day design features an internal driveshaft brake.

On the external version, the handbrake cable attaches to a cam arm on the brake mechanism by a clevis pin. Adjustments are made with the handbrake fully released and the cable disconnected from the cam arm. The arm is moved until the cam is fully released and lies flat.

There should be .025 clearance between the brake linings and the drum on both top and bottom of the driveshaft. At the top there is an anchor and cap screw marking the location where clearance is checked. A flat feeler gauge—the same one used to set ignition points—is used to determine clearance. The clearance can be adjusted by removing the cap screw lock wire and tightening or loosening the screw. At the bottom, there is a lower adjusting nut where clearance is measured and set. Be sure to replace the

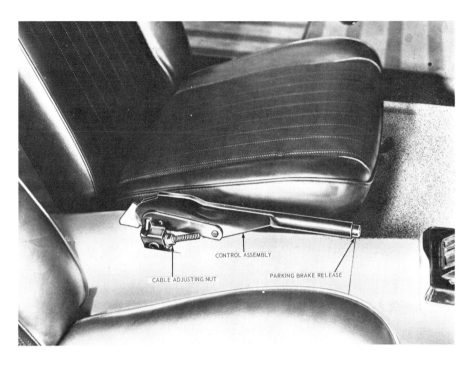

CONTROL ASSEMBLY

CABLE ADJUSTING NUT   PARKING BRAKE RELEASE

A few parking brakes feature cable adjustment at the lever. (Ford)

lock wire on the anchor cap and use a new cotter pin when replacing the brake cable.

Internal driveshaft parking brakes are adjusted with the handbrake fully off and the transmission in neutral position. Before the adjustment can be made, the front end of the shaft must be disconnected so the brake drum rotates freely.

The internal driveshaft brake is much like the regular brake mechanism at each wheel. There is a star-wheel adjuster to regulate the distance between linings and drums. The specs call for at least .010 of an inch lining-to-drum clearance, so check with a feeler gauge. The brake cable length can also be regulated with an adjusting nut, but this should only be done after the lining-to-drum clearance is set correctly.

On a few luxury cars there are tricky devices, such as automatic vacuum release brakes, which may cause you to reach for your shop manual.

THE CLUTCH

On manual transmission cars the clutch is employed to transmit the engine's power to the gearbox. Logically it sits between the rear of the engine and front of the gearbox in its own housing known as the *bell-housing* because of its shape.

The engine's crankshaft turns a heavy wheel known as the *flywheel*. The clutch has a round disc lined with friction-producing material which is pushed into contact with the flywheel whenever the clutch pedal is out. This motion is transmitted back to the gears in the gearbox.

Backing up the abrasive plate of the clutch is a pressure plate filled with little springs. It squeezes the abrasive plate against the flywheel and takes up the shock of engagement and of engine vibration.

Moving the pressure plate and abrasive disc against the flywheel is the clutch *throwout* bearing. It is connected to the clutch pedal by a mechanical linkage (a few cars, mostly foreign ones, use a hydraulic linkage).

The clutch rarely fails suddenly. Usually it begins to slip or jerk and judder. Slipping is often caused by an abrasive plate which is worn or has a

Exploded view of flywheel and clutch assembly. (Pontiac)

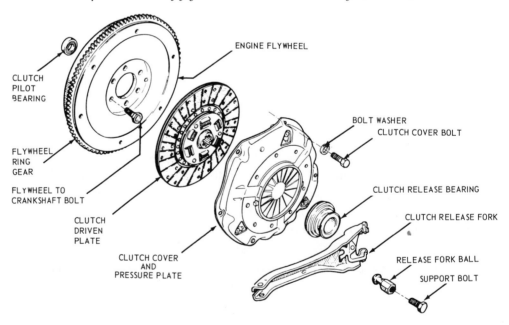

CLUTCH PILOT BEARING

ENGINE FLYWHEEL

FLYWHEEL RING GEAR

FLYWHEEL TO CRANKSHAFT BOLT

CLUTCH DRIVEN PLATE

CLUTCH COVER AND PRESSURE PLATE

BOLT WASHER
CLUTCH COVER BOLT

CLUTCH RELEASE BEARING

CLUTCH RELEASE FORK

RELEASE FORK BALL

SUPPORT BOLT

damaged face (the face material is of asbestos or woven metallic material similar to brake linings). Oil may also be getting on the plate from an engine leak.

Clutch judder may be caused by the disc binding on its input shaft or by a worn pilot bearing. It could also possibly be caused by worn or faulty engine mounts. These parts help control the clutch movement and alignment.

Another cause of either slipping or judder is the clutch linkage being out of adjustment. This is the most common problem and the only one you can fix without taking the engine or gearbox out to get at the clutch mechanism.

A tip-off that the clutch linkage needs adjustment is *free play* at the clutch pedal. The free play is the distance you can press the pedal in before the pedal causes the clutch release lever to contact the throwout bearing. At this point the clutch suddenly becomes much harder to push.

Clutch pedals should have one-half to one inch of free play. More than that and the linkage is out of adjustment. You can measure free play with a ruler.

On all clutches but the diaphragm type the free play is easy to adjust. Diaphragm-type clutches (as used on many GM cars) require a special technique.

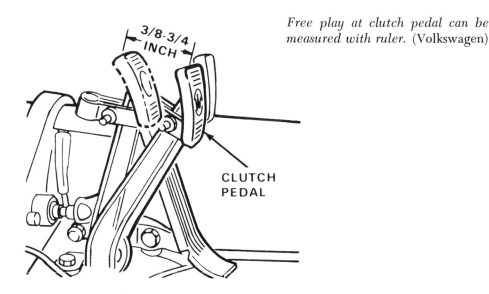

Free play at clutch pedal can be measured with ruler. (Volkswagen)

3/8-3/4 INCH

CLUTCH PEDAL

The mechanically operated clutch linkage can be found by getting under the car and tracing the linkage from the clutch pedal. The linkage is by cable or rods; it can be adjusted by loosening a lock nut, disconnecting a pivot point if necessary, and turning the adjusting nut a few turns. Keep turning the adjusting nut until free play is taken up.

With a diaphragm clutch there is not much plate movement. Adjustments can still be made with a linkage adjusting nut, but the gap between the disc and flywheel must be measured with a feeler gauge and a plate on the bell-housing must come off before you can do it. It is best to check a service manual before you adjust a diaphragm-type clutch.

Hydraulic clutch linkages (found most often on foreign cars) operate with a fluid-filled master cylinder and a slave cylinder much like the ones used with brakes. In fact, some older cars had a combined brake and clutch master cylinder.

With hydraulic linkages there is an adjusting nut between the slave cylinder and the clutch fork. Adjustment is made in a similar manner to mechanical linkages.

On hydraulic linkages it is necessary to check the master cylinder for fluid level and bleed air out of the slave cylinder to be sure that the hydraulic system is operating correctly.

Hydraulic clutch linkages use the same fluid as brakes do and the bleeding process is similar. Crawl under the car and find the bleeder valve on the slave cylinder next to the bell-housing. Crack open the valve one turn with a wrench while someone pumps the clutch. The fluid will come out hissing and spitting until the air in it is exhausted. Then close the valve and fill the master cylinder with new brake fluid to just below the brim.

When linkage is not the problem, you will have to get at the clutch to inspect the plate. Sometimes this can be done through an inspection cover on the bell-housing. Otherwise either the engine or transmission will have to come out to get at the clutch.

Obviously, replacing a clutch is not a simple matter. The technique is not complicated but you will have to do a lot of unbolting to get the engine or transmission out and a hoist plus a heavy floor jack are handy items to have. If you have never replaced a clutch it is best to have the job done by a professional mechanic and watch how he does it.

AUTOMATIC TRANSMISSION CARE

Automatic transmissions are complicated to repair and expensive to replace. All too often they are the first major part to go bad on a car and the car is junked because a new transmission isn't worth the investment.

Many people do not realize that automatic transmissions can be serviced rather easily and that such service may add years to the useful life of a transmission.

Changing the transmission fluid, cleaning the screen or replacing a filter, and possibly installing a new vacuum modulator valve (on transmissions that use this device) should be routine service procedures carried out every 24,000 miles.

To drain the transmission, put the car up on jackstands and crawl underneath looking for the squarish pan which contains the transmission fluid. It should be somewhere behind the similar-looking oil pan.

The transmission fluid pan may or may not have a drain plug. Probably not. The rationale behind this is that, when fluid is changed, the pan should come off anyway so the filter can be changed or the screen cleaned.

Cutaway of Turbo-Hydro transmission. (Pontiac)

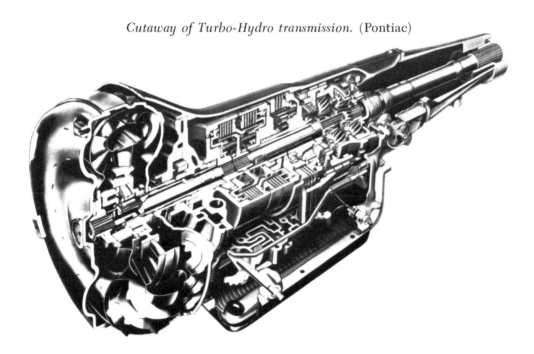

On pans with a drain plug, just loosen the plug to drain transmission fluid. Otherwise you will have to unbolt the pan with the fluid still in it. The technique is to loosen the bolts at the rear of the pan and sides first so the pan tilts backward and most of the fluid falls into a catch basin placed under the rear of the pan. You may have to break the pan gasket loose by prying a bit with a screwdriver if it sticks. It's bound to be messy anyway.

Finish unbolting the pan and take it off, discarding any remaining fluid. Once the pan is off, the fluid filter or screen will be in plain sight. GM and Chrysler Corporation automatics have replaceable filters. American Motors transmissions have a cleanable screen. Ford uses a screen on all transmissions except the C-4, which has a filter.

Purchase the correct filter from an auto parts store at the same time as you buy a new pan gasket. Be sure you specify the make of car and type of transmission when you buy the filter. The surest identification is the number, stamped on a plate attached to the transmission.

Most filters are held on with a few screws, others by clips. Some filters include a rubber ring which connects to an oil tube. This comes off with the old filter and the new one will come with its own ring. Install the new filter the same way the old one came out.

If you have a cleanable screen rather than a filter, remove the screen and clean it in kerosene using a brush. Allow it to dry thoroughly before re-installation.

Clean the drain pan itself with kerosene and scrape the old gasket material from the mating surfaces. Take care in scraping because some transmission housings are made of aluminum, which damages easily.

Thinly coat the gasket with sealer (non-hardening Permatex) on both sides and put it in place. Then replace the pan and tighten the bolts evenly. Tighten one bolt on one side and then one on the other side until all are fairly tight. Then go around and do the final tightening. Put the drain plug back in if there is one.

You may wish to replace the vacuum modulator valve if the transmission has been shifting roughly. This valve uses engine vacuum to control the smoothness of shifting. It is used on GM, Ford, and American Motors cars but not on Chrysler products. On Ford cars, removing the modulator valve requires a special thin wrench, while compact Fords require another type of wrench with a substantial offset. Either type is an inexpensive specialty

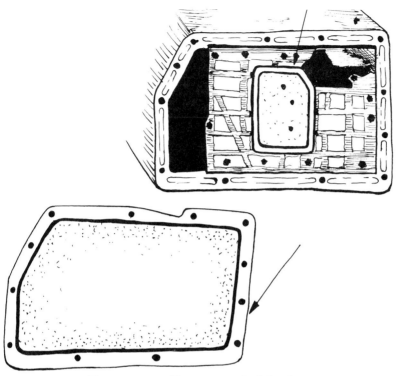

*Transmission filter is reached by lowering
pan. Arrows show cover gasket and filter.*

tool available at most larger parts stores. For the unorthodox, most modula-
tors can also be removed with water pump pliers or vise-grips.

The modulator valve is easy to spot because it has a vacuum hose on its
neck. It looks a bit like a fat spark plug and is screwed into the side of the
transmission. Replace the vacuum hose also if it drips oil or shows signs of
wear.

When you are done below, check your owner's manual for the correct
amount of transmission fluid to put in. Ford products require a different
transmission fluid than other American cars. Fords use Type F fluid, while
all other cars use Dexron. The familiar Type A fluid (Type A, suffix A) is
recommended only for cars produced in 1967 or before. Ford products
1967 or older can use fluid marked F-M2C33-D. The stuff for later cars is
marked F-M2C33-F.

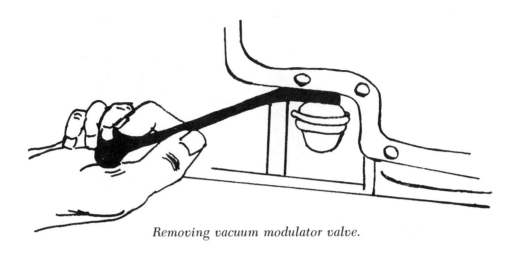

Removing vacuum modulator valve.

Planetary gearset. Arrows show (top to bot-
tom) planetary gear, ring gear and sun gear.

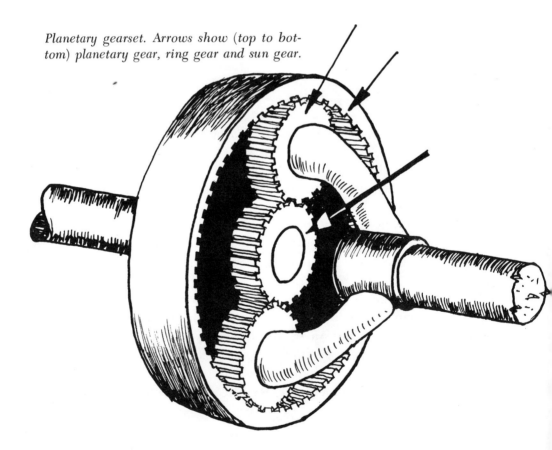

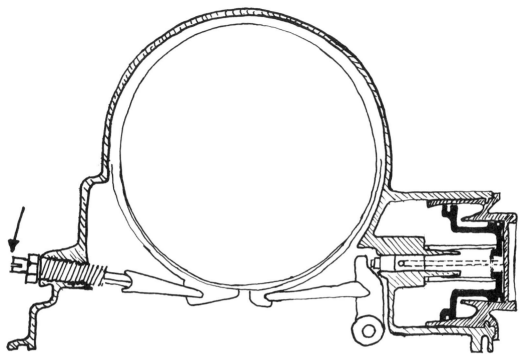

Arrow indicates external band-adjusting screw.

Fill the fluid through the dipstick tube with a flexible funnel. Check fluid level and run the engine while you look underneath to see if there are any leaks. Make a short test run, shifting the transmission through all ranges, and then check the fluid again, adding some if the level has gone down. Each mark on the dipstick is equivalent to one pint.

THE ENGINE: A PHYSICAL EXAM

The internals of the engine are buried away where you can't get at them easily, so engine service tends to be forgotten until the engine begins to perform so badly that a complete overhaul is necessary.

There are ways to prolong the useful life of an engine and they should begin on the day you get a new car.

During the break-in period, car speed should be kept under 60 m.p.h. for the first 500 miles or so and lugging the engine avoided as much as possible. To help seat the piston rings correctly, press your foot all the way down on the accelerator, from time to time, while the car is traveling at a slow speed. Also consult the specific break-in recommendations in your owner's manual.

Some car manufacturers have failed to stress two points which can make a big difference in a car's useful life. One is to use leaded gasoline for the first 1,000 miles of driving and intermittently thereafter. The lead lubricates valves and will guard against the need for a premature valve job. The concept that leaded gasoline causes more pollution has been challenged. Whether it does or not, however, using some leaded gas is essential to valve life according to one major manufacturer (Chrysler) which has conducted studies.

The second precaution is to change oil after only 300 miles or so. This will get rid of the bits and slivers of metal produced in the operation of any brand-new engine.

After the break-in period, change oil at intervals recommended in your owner's manual, or even more frequently if you habitually travel over dusty roads. Use the grade of oil, though not necessarily the brand, recommended by the manufacturer of your car.

Keep the cooling system in good shape to avoid overheating. Repair any oil leaks promptly.

Each time you do a tune-up, you should also use a compression gauge and a vacuum gauge to keep tabs on the condition of the engine.

The compression gauge reads the p.s.i. (pounds per square inch of compression) in each cylinder. Somewhere in the range of 120 to 160 p.s.i. is normal, healthy compression for most cars. In addition, no cylinder should be more than 20 p.s.i. below the rest. You can look up the exact specifications for your car in a service manual.

An inexpensive compression gauge ($5–6) is adequate for a car which has all its spark plug holes in an accessible location. The more expensive gauges come with flex fittings and adapters so they work on V-8s and VWs. On a really high-compression engine (200 p.s.i. or more) it is difficult to hold the rubber probe of the gauge in the spark plug hole against compression, so a gauge with screw-in fittings is preferable.

To use the gauge, begin with a warm engine and remove all the spark plugs. Tag all the plug wires so you can be sure of getting them back in the right order. Place the gauge head in the first spark plug hole and, if it does not screw in, hold it down very firmly.

Now have someone crank the engine with the gas pedal to the floor. Naturally the engine won't start because the plugs are out, but you'll feel the compression in the cylinder pushing against the head of the gauge.

Keep the head of the gauge firmly in the spark plug hole as the engine is cranked over five times. Then check the reading. Do the same for each cylinder.

On a healthy engine each cylinder should be within 20 p.s.i. of the specs. A low reading in all cylinders could mean that the valves are starting to leak or that the piston rings are worn. A variation of more than 10 per cent in one or a few cylinders usually indicates that the piston ring or valves in those cylinders aren't up to snuff. The valves could be worn, burned, sticking, or badly adjusted. Low readings in two adjacent cylinders may mean that the head gasket is damaged between these cylinders.

To pin down the problem further, squirt a little engine oil into the plug holes of the cylinders with a low reading, crank the engine a bit to let the oil do its stuff, then test again with the compression gauge. If the compression jumps considerably, the problem is worn piston rings. The oil temporarily sealed the leaks around each ring and brought the compression up.

Further troubleshooting of an engine with low or uneven compression will require a vacuum gauge. This gauge is also inexpensive and it attaches, via a hose and adapters, to any vacuum outlet coming directly from the intake manifold.

A normal reading for a warm engine at idle is about 17 inches of vacuum. You must correct for altitude, so subtract one inch for every 1,000 feet more than 2,000 feet above sea level.

Blip the throttle once while someone checks the vacuum gauge reading. It should drop way down to 4 or 5 inches, then go up to over 24 before settling down to a normal reading. At idle the pointer should hold fairly steady.

Since a vacuum gauge can indicate a variety of problems, it's fairly tricky to read. A reading of zero or close to it when you blipped the throttle and a slow climb back to normal indicates worn pistons, rings, or wear on the cylinder wall itself.

An occasional drop of 3 to 5 points at idle indicates a sticky, leaking, or burned valve. A steady low reading can be caused by incorrect valve timing (valves need adjustment) or possibly a leak in the manifold or carburetor. Retarded ignition timing will also cause a steady reading lower than normal.

A needle that slowly floats back and forth in the mid-range indicates that the carb mixture is too rich. A fast swing back and forth in the mid-range is due to worn valve guides.

A needle drifting from low to normal and back is indicative of a compression leak between cylinders.

See how the needle reacts when you race the engine at a constant speed. A needle that oscillates rapidly means that the valve springs are weak or even broken.

A normal reading that drops to zero after a while with the engine idling indicates a plugged-up exhaust system.

Tune the engine and adjust the valves, if necessary (the procedure is given in the next chapter). Adjust the carb mixture if you can, or have it adjusted (the problem might only be a clogged-up air filter). Take another vacuum reading and see if problems remain.

By comparing the results from a compression and a vacuum test you should be able to tell what cylinders are not putting out the power and why. You can try to free a sticky valve by using an oil additive made for the purpose (it sometimes works), put a new gasket in a leaky intake manifold, replace a muffler on which the packing has shifted, or know that the rings or valves need replacement.

Worn or broken piston rings let go long enough can lead to the need for a cylinder rebore. Small valve problems can become big ones. Have the work done promptly and you'll pay less.

THE WANKEL ENGINE

Of all the possible replacements for the conventional four-stroke reciprocating engine—electricity, steam, turbines, and all those oddball engines dreamed up by unsung geniuses in Pittsburgh or Peoria and mentioned briefly in *Popular Science* before passing on to oblivion—only the Wankel looks like a winner.

It is a rotary engine invented by Dr. Felix Wankel, first put into production by the German NSU company, and perfected by the Japanese concern of Toyo Kogyo, manufacturers of the Mazda automobile.

The principle of the Wankel engine has been well publicized. Instead of pistons going up and down in cylinders, it has a rotor shaped like a pregnant triangle with bowed-out sides. The rotor revolves in a pinch-waisted "cylinder," or rotor housing, whose shape is known to mathematicians as an epitrochoid.

At each of the three points or apexes of the rotor is a seal contacting the cylinder walls. As the rotor revolves, it always divides the cylinder into chambers where intake of the fuel mix, compression, combustion, and exhaust are going on simultaneously.

In looking at the Wankel engine cycle, you'll note that the rotor makes three revolutions in order to complete a full four-stroke cycle. Thus, following rotor side X as it begins at the end of one exhaust stroke, completing a cycle, you can see it reach maximum volume in the intake chamber, maximum compression (at which point combustion takes place), maximum volume in the exhaust chamber, and then it will return to the point where exhaust is completed.

There are two spark plugs per rotor in the Mazda version of the Wankel, and the trailing plug ignites at ten degrees after the leading one. Two distributors are used for maximum efficiency.

The Mazda Wankel has two rotors, while other applications of the Wankel design have been built with one, three, and four rotors. One of the big advantages of the Wankel is easy adaptation to mass-production of a complete line of engines with varying power by adding on rotors in a modular fashion.

The Mazda Wankels are equipped with fuel, ignition, and electrical systems much like those of any conventional car. One of the few differences in routine service becomes apparent when you try to check the engine compression.

A regular compression tester, inserted in a spark plug hole, would give the highest reading of the three chambers without distinguishing between them. Checking compression requires a recording tester which can graph the peak compression in each chamber. Your Mazda dealer is sure to have one.

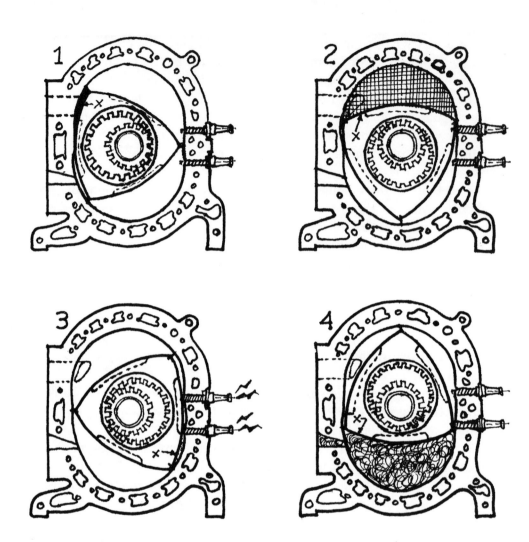

Wankel engine cycle: (1) as rotor revolves, intake begins in one chamber; (2) intake stroke is completed; (3) combustion occurs at maximum compression; (4) exhaust stroke begins.

Don't let the dual distributors and coils snow you. Since the plugs are fired ten degrees apart, each rotor has a leading and trailing plug, fired by the corresponding leading and trailing distributors. The plugs are marked with an "L" or "T" and so are the distributors, so there's little chance of confusion.

Complete tune-up procedures are covered in Chapter 6. Since the Wankel is a little special, however, a few comments are in order here.

Leading and trailing notches on the crankshaft pulley are not marked. To figure out which is which, rotate the engine until one notch is under the pointer. Then the leading notch will be closest to the leading distributor and the trailing one will be nearer the trailing distributor.

Timing is done by hooking a strobe timer to the plug wires of the number one rotor, just as timing is done from the number one cylinder plug wire on conventional engines. With the light hooked to the leading plug wire, you time from the leading notch on the crankshaft pulley and adjust the timing by rotating the leading distributor. Then repeat the procedure for the trailing plug and trailing distributor.

On the newest Wankels there are two sets of points in the leading distributor. The second set retards the spark ten degrees on a cold engine for emission control purposes. When the spark is retarded ten degrees, the dwell angle is also increased by ten degrees. Follow the directions in the factory manual and you won't have any difficulty setting the timing and dwell.

For major engine repairs you'll have to see the nearest factory dealer—at least until Wankels become more common and independent shops begin working on them. The day shouldn't be very far off since Wankels are proving to be reliable and relatively trouble-free—at least in more recent applications. They weigh less for the amount of power developed, have fewer moving parts, and are compatible with the emission control devices necessary to meet increasingly stringent regulations.

What about the car of the future? Perhaps it will have a small Wankel engine inconspicuously tucked away in a little nook to leave space under the hood for 600 pounds or so of safety, convenience, and emission control gadgets.

It takes all the running you can do to keep in the same place, as the Red Queen said to Alice. To get ahead you have to run much faster.

THE BODY BEAUTIFUL

Repairing nicks and dents, or touching up scratched paint on a regular basis, will keep your car from rusting and help you get top-dollar at trade-in time.

Very small nicks in the paint can be covered with touch-up paint available from new car dealers or hardware stores. Use an artist's brush and dab the paint on carefully. This should be done before rust has a chance to form.

Larger nicks or scrapes will require more thorough treatment. You will have to sand, prime, and paint with an aerosol spray to achieve good results.

Use sandpaper with grits ranging from the relatively coarse 220 to the fine 600. For larger areas you will need a rubber sanding block. Remove dirt and wax with a detergent, then sand down to bare metal with coarse grit sanding paper. Always sand in one direction only and not around in circles.

Use an intermediate grit paper for "feathering" the edges around the sanded spot. Wipe the area with a rag to remove loose particles.

Prime with a can of aerosol primer held about a foot from the surface to be sprayed. First use masking tape around the area so stray primer won't get on painted areas. With small surfaces to be painted you can use a sheet of cardboard with a hole cut in it instead of masking tape. Hold the cardboard an inch away from the surface with one hand as you spray with the other.

When you push the nozzle to activate the spray can, or let up on the nozzle, keep the nozzle pointed away from the surface. This will avoid splattering paint which has not been completely atomized by the Freon in the spray can.

Spray on three thin coats of primer allowing ten minutes for drying between coats. Let the primer dry for an hour and then wet-sand it. Wet-sanding is accomplished by keeping a trickle of water from a hose playing over the area as you sand it. This can also be accomplished with a basin of water and large sponge which is frequently squeezed above the area. Use intermediate-to-smooth grit wet sandpaper starting with a slightly coarse grit to smooth the primer and ending with a finer grit to blend the primer with the surrounding paint.

Let dry overnight, then paint with the final coats of color. Mask again and use the same painting techniques as used in applying the primer. Apply paint very lightly in three or four successive coats with twenty minutes for drying between coats.

Let the paint dry overnight again. Finally use abrasive rubbing compound to get a high-gloss on the paint and blend it in further. Don't rub too hard or you might rub away your beautiful new paint job.

DENTS AND HOLES

You can repair dents and holes—even fairly large rusted-out areas—with a combination of techniques which include pounding out, patching with body filler, and using fiberglass cloth and epoxy for holes.

Advanced body work takes real skill and special equipment, so you can't expect to do a good job on panels that have been thoroughly bashed. Confine yourself to the less ambitious jobs and you'll be more satisfied with the results.

A substantial hammer and the contoured iron "dolly" used by body men are the essential tools. If you must remove inside door panels a special tool to take off interior door handles is needed. The pin-type puller works on most American cars with the exception of GM products and Fords prior to 1962. These require a clip-type puller.

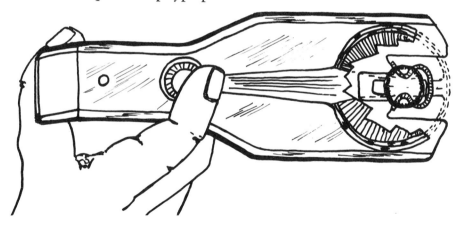

Special tool for door handles or window cranks held on with clips.

Some shallow dents will come out by just pushing from the inside until they snap back to the original position. Most require pounding with the hammer and dolly from the inside. This will probably cause the dent to bulge outward and it must then be flattened by pounding from the outside. The technique is to keep working on the inside and outside alternately until the surface is as smooth as possible.

Small dents or those which will not come smooth with pounding can be patched with plastic body filler. This is done by sanding with medium-fine grit paper (no. 320 or 360) until the metal is bared. Clean away the sanding residue and apply plastic body filler to the dent. Smooth it with the edge of a piece of cardboard and let it overlap the sides of the dent a bit. Allow filler to dry for a few hours.

Sand the filler with coarse sandpaper and then medium grit until it is perfectly smooth. Then prime and paint.

Holes can be patched with fiberglass cloth and epoxy kits available at some hardware stores and at many boat supply stores.

Start by sanding down to bare metal a half-inch or so all around the hole. Wipe away residue and spray primer on this area. Cut fiberglass cloth to overlap the hole a half-inch on all sides and impregnate the cloth with epoxy resin. Apply it to the hole. Use two layers of cloth for larger holes (above one and one-half inches in diameter) and allow thirty minutes between layers for the epoxy to partially set. In areas colder than room temperature the epoxy will not set unless a heat lamp is used on it. Allow to dry overnight, then sand with coarse grit paper and medium-fine grit until smooth. Finish with primer and paint.

Large rusted out areas can be repaired with the use of a back-up inside sheet of cardboard temporarily affixed with tape. A number of layers of cloth may be necessary to provide the needed strength. The technique has definite limits so you are better advised to look for a whole new panel in a junkyard.

Rusted out areas in the floor or trunk compartment can be repaired by pop-riveting sheet metal over the defective area. Use cardboard for a template if necessary and then cut the sheet metal to size with tin snips. Be sure the repair overlaps so that you attach to unaffected metal on all sides.

Drill holes where the pop rivets will go, then use a rivet gun to set them. Sheet metal screws can also be used, although rivets are faster to work with and hold better.

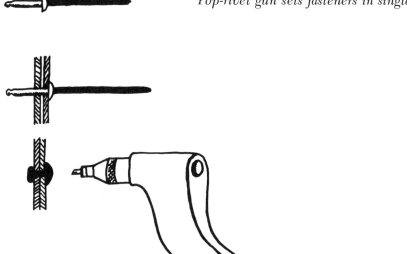

Pop-rivet gun sets fasteners in single stroke.

THEFT PROOFING

The only trouble with touching up the paint job on your car and other cosmetic treatment is that you have not only made the car more attractive, but also more susceptible to thievery. So why not combine a little theft-proofing with a general shape-up?

Only about one out of every hundred cars is stolen each year, but don't count on these odds applying to you. Should you happen to own a late-model car of a popular make, and regularly park it on the street in a metropolitan area, the odds are considerably greater. In some neighborhoods they approach fifty-fifty.

Four-fifths of all cars stolen are taken by non-professionals who are under twenty-one years old. Just the basic precaution of taking the key out of the ignition and locking the car will discourage most casual thefts. Protection against the pro or semi-pro will require more effort.

There are all sorts of burglar alarm and ignition locking devices, some of which cost well over $100. None of these will make your car absolutely theft proof since some professionals keep abreast of all the latest developments in anti-theft gadgetry and take some pride in being able to deactivate more complex devices. There is even a tale about a legendary specialist who only steals expensive burglar alarms.

Although there is no ultimate anti-theft system, there are inexpensive methods for raising the odds against thievery which work just about as well as the more complex alarms and locks. A sturdy hood lock combined with any device designed to make starting the car impossible without under-hood access can provide a high degree of security.

Cylinder-style hood locks are convenient to install and use, but they can be picked far too easily. A clumsier but more secure method is to loop a piece of chain between an attaching point on the underside of the hood (such as the cross bar of the latch mechanism) and a grille or bumper component which cannot be unbolted easily. Use half-inch hardened chain which is resistant to all bolt cutters in common use, and run the chain through a piece of hose to protect the car's finish. Buy the best padlock you can afford to lock the chain in place. Seek a locksmith's advice if in doubt about the most secure padlock.

There are many ways to prevent starting the car without opening the hood. At no cost, one method is to make it a practice to remove the distributor rotor when you park the car; or take off a battery terminal clamp, plug up the fuel line, block the points with a small piece of cardboard, or pull off the thick wire between the coil and distributor.

If the above methods seem too crude or time consuming, you can purchase a battery terminal cut-off switch, a simple in-line electrical switch to install in one of the ignition circuit wires, or a needle valve and fittings which will block off your fuel line with just the turn of a handle. Auto supply and hardware stores can usually furnish the fitting you need for a few dollars.

A few inexpensive commercial devices can supplement your own ingenuity. Locking bars which fit between the steering wheel of a car and the brake pedal or shift lever are reasonably effective. To prevent your good radials and custom wheels from being lifted by the boys from midnight parts requisition, it may pay you to spend about eight bucks and install locks which replace a lug nut on each tire. (Just be sure you have the key along when you get a flat.) For a few more dollars you can buy straight-sided door lock knobs which replace the mushroom-shaped locking buttons on most car doors. They effectively prevent adolescents with a length of bent coat-hanger wire from fishing the button up and getting in to steal your tape deck.

On second thought, buy only the type of tape deck which can be slipped out of its mounting and taken inside. Tape decks, other valuables, and even packages which contain nothing valuable but look like they could, should not be left in view. Lock everything that you can't take with you in the trunk compartment.

Putting a sticker on your windshield saying that the car is protected by a burglar alarm is inadvisable since some thieves may take it as a tip-off that you carry valuables in the trunk. A better idea is just to park in well-lighted areas along busy streets, whenever possible.

Every car has a vehicle identification number plate in an accessible location such as on top of the dash near the window or on the door in the vicinity of the latch. Consult your owner's manual for location of the VIN (vehicle identification number) and keep a record of it in your wallet. While it is easy enough for anyone who swings with your sweet chariot to remove the VIN plate, the number is also marked somewhere else on the car in a secret position established by the manufacturer and unknown to casual thieves and joy riders. When you report your car stolen, the first thing the police will ask for is the VIN.

STORAGE

A few precautions will keep your car serviceable when you store it for more than a month or two without someone around to run it periodically. Start with a good wash and wax job, being sure to hose down the chassis and running gear in order to remove road salt and built-up mud and grit.

Remove tools and spare tire from the trunk compartment. Pull up the trunk compartment mat and see that the trunk is dry. Then remove the rubber drain plug at the bottom of each door and allow any water in the door to run out.

Go around the car with a can of aerosol silicone lube and spray scrapes in the body sheet metal, door and trunk latches, and other areas subject to rust.

Drive the car to your storage place. Jack it up, place jackstands or other sturdy supports at the jacking points beneath the car, and take the wheels off. Store the tires off the ground and away from damp walls.

Remove the battery and store it in a cool place. Drain the radiator and engine block and drain off the oil (you'll be replacing it with fresh stuff when you put the car back in service). Drain or siphon off the gasoline.

Remove each spark plug, squirt some penetrating oil into the cylinder, and replace the plug. If you are worried about the plugs corroding in place, which can happen during long storage in a humid area, put Never-Seez or a similar lube on the plug threads.

Try not to store your car in an old barn where rats and insects are sure to claim nesting privileges. Also avoid cellars that flood.

6

How to Do a Tune-Up

● In the broadest sense a tune-up means adjusting and replacing minor parts in the ignition system, fuel supply system, and emission control system in order to make the engine run more smoothly and develop its full power. Adjustments can be made to the engine itself also.

Since the fuel supply and emission control systems have been covered previously, we will stick to ignition work and valve lash adjustment in this chapter.

To understand the steps in a tune-up you should know how each part in the ignition system functions. You have already read a simplified explanation in Chapter 1. Here we will go into the ignition system more deeply.

HOW IT WORKS

The spark needed to ignite the fuel mixture in the engine cylinders is produced when high voltage jumps the gap between the electrodes in a spark plug.

The high voltage originates in the coil. The coil creates this current by stepping up low voltage which comes through a thin wire from the battery circuit via the ignition switch. Another thin wire attached to the coil comes from the point mechanism in the distributor.

Although the coil receives current from the battery, the electric circuit is not completed until the distributor points close. When the points come together, the coil circuit is completed and current flows through.

The points are like an on-off switch which makes and breaks electrical contact; they must open and close many times a second to start the cycle which leads to a spark igniting the mixture in each cylinder.

The coil, receiving low voltage whenever the points close, converts the current to high voltage and sends it back to the top part of the distributor through a thick wire. The high voltage passes through the distributor condenser (think of the condenser as a filter screening out the harmful part of the voltage—this explanation is not technically correct but the basic concept is valid) and goes to the distributor rotor. As the rotor revolves it makes contact with the terminal of each spark plug wire in the distributor cap. High voltage goes to each plug in turn and the spark is created to fire the mixture in the cylinders.

The rotor turns because it is connected by gears to the engine camshaft.

Cross-section of coil. (Chevrolet)

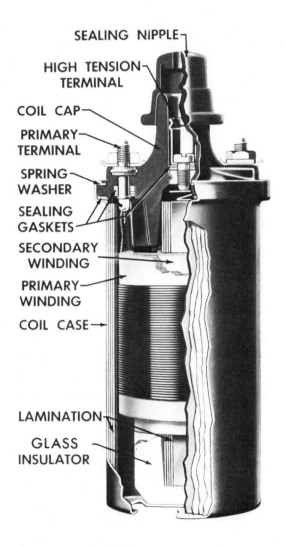

SEALING NIPPLE

HIGH TENSION TERMINAL

COIL CAP

PRIMARY TERMINAL

SPRING WASHER

SEALING GASKETS

SECONDARY WINDING

PRIMARY WINDING

COIL CASE

LAMINATION

GLASS INSULATOR

At the base of the rotor drive is a cam which looks like a large nut with four, six, or eight sides, depending upon the number of cylinders. The points or lobes of the cam push against a fiber rubbing block connected to the moveable point arm. As each high point or lobe of the cam pushes against the rubbing block, a spring on the point arm is stretched and the points open. When the lobe moves past the spring contacts, the points close. There are as many lobes on the cam as there are cylinders in the engine. Each time the points close voltage is created and sent to one cylinder to create a spark.

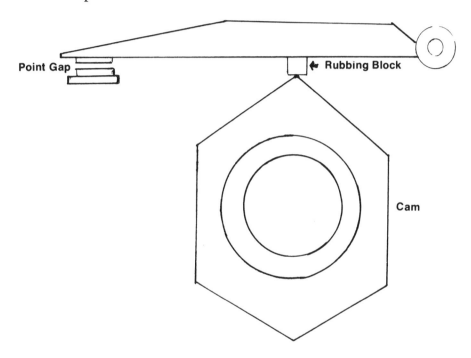

Points open when rubbing block is at cam apex.

In order for everything to work right, the gap between the spark plug electrodes must be just wide enough to create a strong spark, and the ignition points must open to the exact point necessary to break the circuit. If the points open too wide they could not close fast enough to create the spark in each cylinder at the right time.

Along with the spark plug gap and point gap, the third critical factor in the ignition system is timing. The spark must arrive in each cylinder when the piston is in its combustion cycle and has compressed the gas/air mixture which is then ignited by the spark.

If you could look into the engine and see when a particular piston was at the top of its combustion stroke, and if you could look at the spark plug at the same time to see if it was firing, then you would know the timing was right. The relationship between the cylinders remains constant; when the timing is correct for the number one cylinder, it is correct for all of them (unless the internal timing between crankshaft and camshaft is messed up —a major problem we won't go into here).

Since you can't look into the engine without taking it apart, the timing must be measured indirectly. The engine crankshaft is connected to the pistons and rotates when it is "pushed" by them. This rotation can be measured by the degrees in a circle. By determining the degree the crankshaft should be at when a particular cylinder is ready to fire, the timing of the ignition system can be checked.

The crankshaft has a "wheel" at either end. At the rear it is the flywheel which mates to the clutch and transmits power. In front it is the circular pulley which drives the fan belt. A timing mark can be placed on either wheel. When this mark is aligned with a stationary mark the number one cylinder is ready to fire.

Tune-up manuals specify where the timing mark is placed. It may be at top dead center (meaning the point at which the piston in the number one cylinder reaches its zenith), or so and so many degrees before or after top dead center (TDC). All you have to know is which timing mark to use and where it is placed.

Since it is difficult to tell when the spark plug is firing, you must use an electrical device to determine when high voltage is coming through the spark plug wire. This can be done by attaching a timing light to the wire going to the number one spark plug. Whenever voltage comes through the light will flash. It will flash off and on rapidly like a stroboscope.

Now all you have to do is point your timing light at the timing marks and turn on the engine. If the timing marks are lined up whenever the light flashes on, then the timing is correct. This is not the only way to time the ignition system, but it is the most common one.

The timing is checked with the engine idling. However it is necessary that the timing change when the engine is speeded up. At low engine speeds the spark plug must fire before the top of the compression stroke in order to get the most power. At higher speeds it should fire a bit later for maximum power. This is due to the lag time in the combustion process itself. When the pistons are moving more slowly, at low engine speeds, the combustion takes place as the piston reaches top dead center due to the lag time. When the pistons move faster the spark must be fired sooner to allow combustion to take place as the piston reaches TDC.

Earlier cars had a manual spark advance mechanism. As the engine was speeded up the spark was advanced (made to fire sooner) by a lever on the dashboard or steering column. On modern cars the spark is advanced automatically.

Most cars have two separate automatic spark advance mechanisms. One works by spring-loaded weights which separate by centrifugal force as the engine speeds up. This is known as the centrifugal advance, naturally. The other is controlled by manifold vacuum and called the vacuum advance. It has a spring-loaded mechanism controlled by a vacuum diaphragm. The more vacuum there is, the more the diaphragm pushes against the spring and causes the spark to be advanced.

A few cars have been made with only one type of advance mechanism; either the centrifugal or vacuum variety. However, the centrifugal type alone offers poor fuel economy at low speeds and may advance the spark excessively at high speeds. The vacuum type alone limits acceleration and top speed since it does not advance the spark enough when the engine is under load and vacuum in the manifold is low. The two advance mechanisms generally work best as a team.

PREPARING FOR THE JOB

Now that you have a basic understanding of how ignition systems function, we can begin the tune-up. Get the parts you'll need at an auto supply store. Purchase spark plugs (one per cylinder), a set of points, condenser, distributor rotor, and distributor cap.

It may not be necessary to replace the distributor rotor and cap nor the condenser, but you should keep these items on hand in case you need

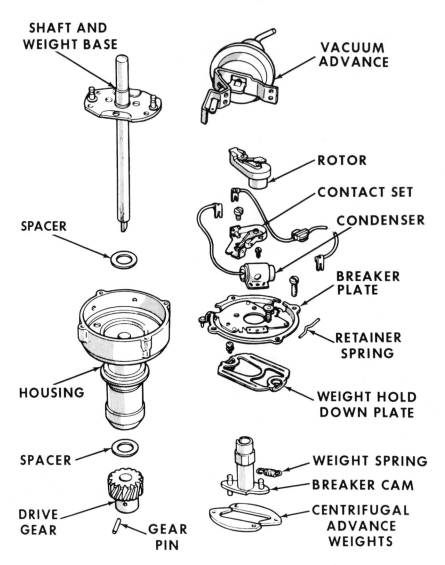

SHAFT AND
WEIGHT BASE

VACUUM
ADVANCE

SPACER

ROTOR

CONTACT SET

CONDENSER

BREAKER
PLATE

RETAINER
SPRING

HOUSING

WEIGHT HOLD
DOWN PLATE

SPACER

WEIGHT SPRING

BREAKER CAM

DRIVE
GEAR

GEAR
PIN

CENTRIFUGAL
ADVANCE
WEIGHTS

Exploded view of distributor. (Pontiac)

them. Add any parts not used to your emergency road kit (wrapped in foam to avoid breakage) and carry them with you in your car.

Each manufacturer has a different code for the same plugs. The clerk in the auto store should have a chart listing the correct plugs, in his brand, for your car.

Points, condenser, and possibly a rotor will be sold in a packaged kit which may also contain a little capsule of distributor cam grease. Printed on the package will be the model and year of car it is designed for.

The distributor cam grease will be enough to grease three or four cams, since it should be applied very lightly. If there is no little capsule in your kit, you can get a tube of the stuff at most auto supply stores. Don't use any other kind of grease since it will break down quickly under the heat generated.

The distributor cap is usually sold separately. The model, engine, and year of car it fits will also be printed on the package.

Your total expenditure should be under $9 for a six-cylinder car and under $12 for an eight-cylinder. Compare this with the $20 to $40 that most garages will charge for a tune-up and you can see that there is a worthwhile saving in doing it yourself.

Some garages do advertise a "complete" tune-up for $9.95 or some such figure. This is a come-on since you will be charged extra for parts and the mechanic will probably try to sell you repairs you don't need. The $9.95 tune-up can wind up being the most expensive of all.

STEPS IN A TUNE-UP

Assemble your tools: a spark plug wrench (ratchet wrench with a deep socket sized $^{13}/_{16}''$, or $^{5}/_{8}''$ on a few cars), ignition wrenches (small open-end wrenches), flat and wire feeler gauges, small screwdrivers, and a timing light. Your wire feeler gauge, designed for use with spark plugs, should have a small file and an electrode bender attached to it. If you want to file the points instead of replacing them, you will need a plasticized file (brand name: FlexStone) which looks like an emery board. An emery board itself is too coarse to give the points a mirror-smooth finish. Some cars will also require that you use a special wrench to reach the distributor lock nut and loosen it.

The steps in a tune-up are:
1. Removing and gapping the spark plugs.
2. Removing the distributor from the engine (if necessary).
3. Taking out the rotor, condenser, and points.

4. Cleaning the breaker plate and centrifugal advance mechanism.
5. Putting in the new condenser and points or re-installing the old parts if they are still good.
6. Setting the point gap.
7. Replacing the distributor (if you took it out) and putting the rotor back on.
8. Replacing the distributor cap.
9. Installing the new spark plugs.
10. Timing the ignition system.
11. Adjusting valve lash (on cars where this is a necessary procedure).

It all sounds tougher than it really is. Once you have done it a few times and have the steps down, the whole job can be done in an hour or less.

REMOVING THE SPARK PLUGS

Remove all spark plug leads and tag them so you can be sure of replacing them in the right order. On a four- or six-cylinder engine you can tag the leads by number from front to rear. On a V-8 there are two blocks of cylinders so the leads should be tagged A and B with numbers from front to rear.

Remove the spark plug leads by grasping them where they fit over the plug ends and pulling them off gently but firmly. Plug wires which are frayed or cracked should be replaced with new wires and end boots. Otherwise they will "leak" electricity and cause hard starting along with loss of engine power due to a weak spark. Tag each plug lead with a piece of tape marked with the number of the cylinder it goes to.

The plug wires which are original equipment on most cars are not of high quality. Original equipment wires should be replaced after 36,000 miles or less. Get a good brand of resistance wire, such as Belden or Packard, and the screw-on type of end connectors rather than those that punch through the wire's insulation. You can buy a set of wires custom fit for your car or cut the wires and put end connectors on yourself. Good spark plug leads can last the life of your car and be highly resistant to grease and moisture.

Incidentally, you should avoid the metallic-core wires which are popular on hot rods. They conduct electricity fine but they also make the radio in

CARBON FOULED

IDENTIFIED BY BLACK, DRY FLUFFY CARBON DEPOSITS ON INSULATOR TIPS, EXPOSED SHELL SURFACES AND ELECTRODES. CAUSED BY TOO COLD A PLUG, WEAK IGNITION, DIRTY AIR CLEANER, DEFECTIVE FUEL PUMP, TOO RICH A FUEL MIXTURE, IMPROPERLY OPERATING HEAT RISER OR EXCESSIVE IDLING. CAN BE CLEANED.

OIL FOULED

IDENTIFIED BY WET BLACK DEPOSITS ON THE INSULATOR SHELL BORE ELECTRODES CAUSED BY EXCESSIVE OIL ENTERING COMBUSTION CHAMBER THROUGH WORN RINGS AND PISTONS, EXCESSIVE CLEARANCE BETWEEN VALVE GUIDES AND STEMS, OR WORN OR LOOSE BEARINGS. CAN BE CLEANED IF ENGINE IS NOT REPAIRED, USE A HOTTER PLUG.

GAP BRIDGED

IDENTIFIED BY DEPOSIT BUILD-UP CLOSING GAP BETWEEN ELECTRODES. CAUSED BY OIL OR CARBON FOULING. IF DEPOSITS ARE NOT EXCESSIVE THE PLUG CAN BE CLEANED.

LEAD FOULED

IDENTIFIED BY DARK GRAY, BLACK, YELLOW OR TAN DEPOSITS OR A FUSED GLAZED COATING ON THE INSULATOR TIP CAUSED BY HIGHLY LEADED GASOLINE. CAN BE CLEANED.

NORMAL

IDENTIFIED BY LIGHT TAN OR GRAY DEPOSITS ON THE FIRING TIP. CAN BE CLEANED.

WORN

IDENTIFIED BY SEVERELY ERODED OR WORN ELECTRODES CAUSED BY NORMAL WEAR SHOULD BE REPLACED.

FUSED SPOT DEPOSIT

IDENTIFIED BY MELTED OR SPOTTY DEPOSITS RESEMBLING BUBBLES OR BLISTERS. CAUSED BY SUDDEN ACCELERATION. CAN BE CLEANED.

OVERHEATING

IDENTIFIED BY A WHITE OR LIGHT GRAY INSULATOR WITH SMALL BLACK OR GRAY BROWN SPOTS AND WITH BLUISH-BURNT APPEARANCE OF ELECTRODES, CAUSED BY ENGINE OVERHEATING, WRONG TYPE OF FUEL, LOOSE SPARK PLUGS, TOO HOT A PLUG, LOW FUEL PUMP PRESSURE OR INCORRECT IGNITION TIMING. REPLACE THE PLUG.

PRE-IGNITION

IDENTIFIED BY MELTED ELECTRODES AND POSSIBLY BLISTERED INSULATOR. METALLIC DEPOSITS ON INSULATOR INDICATE ENGINE DAMAGE. CAUSED BY WRONG TYPE OF FUEL, INCORRECT IGNITION TIMING OR ADVANCE, TOO HOT A PLUG, BURNT VALVES OR ENGINE OVERHEATING. REPLACE THE PLUG.

When you remove spark plugs, check their condition against this chart. (Ford) your car and the neighborhood TV sets go haywire.

With the plug wires detached at the spark plug end and neatly tagged, you can now remove all the plugs using your ratchet wrench and plug socket. You may need an extension or swiveling adapter on your socket handle to reach the plugs. Most four-cylinder cars and sixes are easy. Some V-8's loaded with accessories have plugs that require a lot of ingenuity to reach.

Wipe off dirt or oil around each plug hole with a rag soaked in solvent. Don't let dirt go down the plug hole where it will get into the cylinder and cause mischief.

Inspect the plugs carefully. A normal plug that has been run for a while will have a tan or grayish deposit on the insulator and electrodes. The deposit should be fairly light. A thick, crusty deposit indicates that you have been doing a lot of low-speed driving. Oil deposits on the plugs mean that there is probably an internal oil leak in the engine caused by worn rings or valve guides. Carbon deposits are due to an overly rich mixture or an ignition system which has not been putting out enough voltage.

Burned electrodes are caused by a spark which is advanced too much, poor ignition timing, or the use of gas with too low an octane rating. Burned and carbon-coated electrodes mean that there are heavy carbon deposits in the cylinders which must eventually be cleaned out.

When the spark plugs are abnormal, tune the engine and test compression and vacuum. If the engine seems to run well, try having the fuel mixture adjusted by a mechanic who uses an exhaust gas analyzer. He may also advise you to switch brands of gasoline and change your driving habits slightly.

You must determine whether to replace or just clean and regap the plugs. On most cars the plugs will last 12,000 miles if they are cleaned at the halfway mark. After 12,000 miles the plugs should be replaced even if the electrodes are still in fair shape. Running plugs too long will hurt a car's performance and gas mileage; ultimately the engine may be damaged.

The easy way to clean plugs is to sand blast them in a machine which many service stations have. After sand blasting, the center electrode must be filed flat. If you do not have access to a sand blasting machine, you can scrape the deposits off with a wire brush and then file. Bend the side electrode out of the way while you file the center one flat.

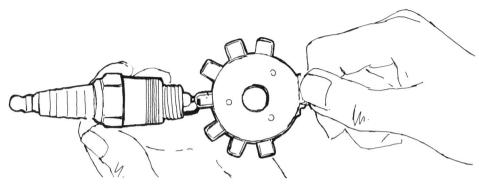

Checking spark plug gap.

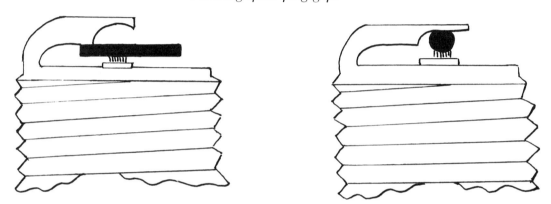

Flat feeler gauge blade will not show correct gap (left); wire gauge fits notch in electrode (exaggerated here) and gives true reading.

Use your wire gap gauge to regap the plug. If you do not know the correct gap, check your owner's manual or service manual. On cars manufactured after 1968, tune-up data is given on a plate or decal somewhere in the engine compartment. You can also send $1 to J. C. Whitney (see Appendix C for address) and get the latest edition of the *Complete Engine Tune-up Guide*, which gives the spark gap, point gap, timing degree, and other specifications for American cars and many popular foreign makes.

Use the correct-sized wire feeler on your gapping gauge. It should just fit in between the electrodes with a slight drag when removed. The reason it is necessary to use a wire gauge rather than a flat one is that there is a slight notch under the side electrode. The wire feeler goes into this notch.

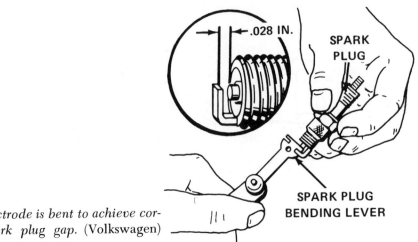

Side electrode is bent to achieve cor-*rect spark plug gap.* (Volkswagen)

If the gap is incorrect, use the electrode bender on your gapping tool to bend the side electrode up or down. Do not tap the electrode with a hammer or use any other crude method which will damage the plug.

Some plugs have a metal gasket, while others have a tapered seat and require no gasket. Where gaskets are used, the new plug should come with a replacement gasket.

On brand-new plugs you will still have to check the gap. The gap is not set at the factory to fit any particular car.

Do not replace the plugs immediately. You will want to leave the plugs out while checking the point gap since the engine is easier to turn manually when the plugs are removed.

REMOVING THE DISTRIBUTOR

You have a choice between taking the distributor out of the car or installing new distributor parts and setting the point gap without removing the distributor. It is easier to leave the distributor in if you can get at it to work, but you can do a more thorough job by taking it out.

Whether you remove the distributor or not, the first two steps are the same. First remove the distributor cap by flipping off the wire clips on each side with a screwdriver or loosening screws at each side. Note how the cap fits on so you don't replace it backwards. Leave the spark plug wires attached to the cap to save the trouble of tagging them at the distributor end. Put the cap with the wires attached to it off to one side.

Then look at the vacuum advance mechanism which is attached to one side of the distributor and shaped like a mushroom head. Remove the vacuum line running to it and pinch the line off at its end or else plug it up.

To remove the distributor you will need a wrench to reach the lock nut. On some V-8 cars the lock nut is inaccessible by ordinary means so you will need a special wrench with a long bent handle to reach it. A distributor lock nut wrench is a specialty tool sold by professional parts stores. Get one designed for your car if you need it.

Before removing the distributor, mark its position by placing a piece of tape on the distributor body and another piece on the hold-down mechanism. Since the timing changes when the distributor body rotates, you must place the distributor body back in the same position it was in when you took it off.

Use another bit of tape to mark the position the rotor tip points to on the distributor body. You will want to replace it in the same position on the distributor when you put everything back.

Loosen the distributor lock nut and pull the distributor body upward, out of engagement with its drive mechanism. If the rotor turns a bit as you pull it off (it will do this on certain types of distributor drive mechanisms), don't worry about it. Merely remove the tape marking the rotor position and put it back on the distributor body at the final position the rotor tip points to.

REMOVING DISTRIBUTOR PARTS

Most rotors just lift off, although some (General Motors V-8 cars, for instance) attach by screws on either side. If the rotor is cracked or has carbon tracks on it, then it must be replaced. A burned rotor tip is also cause for replacement, but slight deposits on the tip can be removed by burnishing with a FlexStone.

Lubricate the cam under the rotor with a very light coat of distributor cam grease. If there's a felt wick in the center of the cam, put a few drops of oil on it.

Rotor on Delco distributors (top) is fastened with screws (see arrows); others, as shown below, just lift off.

The next step is to clean the points or replace them with a new set. To replace the points, take off the primary and ground wires and unscrew the attachment to the breaker plate. Note how everything came off so you can fit it back the same way.

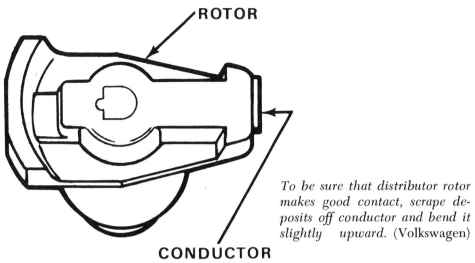

ROTOR

CONDUCTOR

To be sure that distributor rotor makes good contact, scrape deposits off conductor and bend it slightly upward. (Volkswagen)

Points which are not badly pitted or burned can be used again. Dress the points to a mirror finish with the FlexStone and be sure they are flat.

Also ensure that the point faces are properly aligned. If the contacts are not flat against each other, bend the stationary point arm with a pair of needle-nose pliers (or a point bending tool if you have one) until alignment is correct. Never bend the moveable arm.

Some distributors have dual sets of points and both must be burnished or replaced with new ones.

CLEANING THE BREAKER PLATE AND ADVANCE

Before putting in the new distributor parts, remove the breaker plate at the bottom of the distributor and clean it thoroughly using solvent. Also clean the weights and linkage of the centrifugal advance mechanism. Put a few drops of oil on the linkage.

REPLACING DISTRIBUTOR PARTS

Attach the points by screwing them on and replacing the primary and ground wires in the same way they came off.

Replace the condenser by putting the new one in the hold-down strap and tightening the screw on the strap. Be sure the screw is tight since this is the condenser's ground. Although condensers can last 25,000 miles or more, most people replace them as a general precaution since they're cheap.

Put the rotor back on after setting the point gap.

SETTING THE POINTS

To measure the point gap, the points must be fully open. If the distributor is off the car all you have to do is rotate the cam until one of the lobes is against the little projection (rubbing block) on the point arm.

When you have not removed the distributor it is a little more difficult. Have someone crank the engine for a second until the lobe is contacting the rubbing block or close to it. You may have to try a few times before the position is close. Then put the rotor on temporarily and use it as a handle to turn the drive mechanism until the cam lobe is against the rubbing block holding the points open to maximum gap. You can also turn the drive mechanism by tugging on the fan belt. However, this puts too much strain on the belt and is not recommended. On Volkswagens and some other cars you can turn the engine by getting a wrench on the nut in the middle of the generator pulley. On some cars the easiest method may be to push the car forward a bit until the rubbing block is at the high point of the cam.

With the points fully open, check the gap with a flat feeler gauge which is the correct thickness. Check your owner's manual or tune-up manual to find the gap specified. Make sure you hold the feeler blade perpendicular to the points when checking the gap. At the correct gap there should be a slight drag on the blade as you remove it. If the gap is incorrect, change it by turning the point adjusting screw. Tighten the screw and check the gap again. On some distributors there is a lock screw and adjusting screw controlling the point gap, while others have a lock screw and slot which allows the position of the point arm to be adjusted.

Delco-Remy distributors on late-model General Motors' cars and AMC V-8's have a different method of point adjustment. These distributors have a little window on the exterior of the distributor body which opens up to allow point adjustment with the engine running. With the distributor fully assembled and the plug leads connected to the plugs, the engine is started.

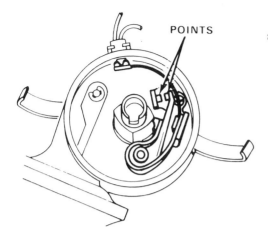

POINTS

Points must be fully open when gap is checked. (Volkswagen)

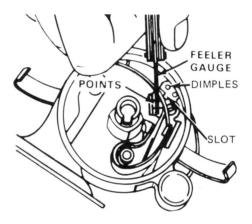

FEELER GAUGE

DIMPLES

POINTS

SLOT

Feeler gauge blade should be held perpendicular to points when measuring gap. (Volkswagen)

A little hexagonal Allen wrench (which usually comes in the points kit for these cars) turns the adjusting screw inside the window. The procedure is to turn the screw clockwise until the engine just begins to miss, then back off one-half turn. Or use a dwell meter for a more accurate setting.

Dual-point distributors, which have been primarily used on performance cars, have now been adopted by a few economy cars (such as the Datsun 510 and Chevvy LUV pickup truck) to aid emission controls. Both sets of points must be gapped on these distributors.

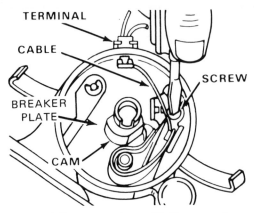

Adjusting points. (Volkswagen)

Adjusting points on Delco distributor. (Pontiac)

REPLACING THE DISTRIBUTOR AND ROTOR

Put the distributor body back on and press it down to engage the gears. If the gears engage but the distributor body won't go down that final bit, the oil pump drive is probably not aligned right. Stick a long screwdriver down the distributor hole to move the drive out of the way and then try putting the distributor on again. It should go down all the way. Rotate the body so that the pieces of tape you used to mark the correct position line up. Then tighten the lock nut.

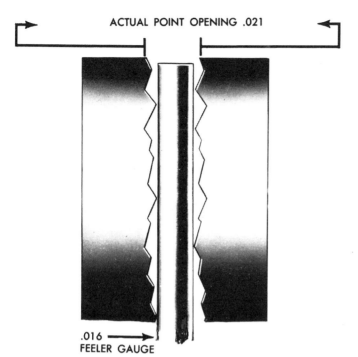

ACTUAL POINT OPENING .021

.016 →
FEELER GAUGE

Point surface becomes irregular with wear. Gap reading, as shown, will be inaccurate unless points are first filed smooth. (Chevrolet)

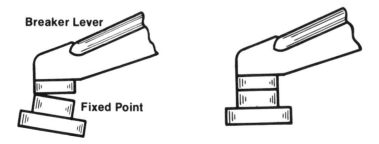

Breaker Lever

Fixed Point

You can correct lateral misalignment (shown at left) by bending fixed contact support. Never bend movable arm (breaker lever). Diagram at right shows proper lateral alignment. (Chevrolet)

Replace the rotor by aligning the notches on either side with the notches in the drive mechanism. The rotor tip should line up with the piece of tape you put on the distributor body to mark its position. Be sure it is seated firmly.

CHECKING THE VACUUM ADVANCE

Before replacing the distributor cap, check the operation of the vacuum advance mechanism by moving the breaker plate forward by hand as far as it will go. This is the full advance position. Place your finger over the vacuum port (where the hose attaches) and release the breaker plate. It should move only a little bit and then hold in the advanced position.

When you take your finger off the vacuum port, the breaker plate should snap back to its original position.

If the breaker plate did not hold in the advanced position, vacuum is leaking and you need a new vacuum advance unit. If the plate didn't snap back, the mechanism is binding and you didn't do your cleaning job thoroughly. It is also possible that the spring on the vacuum advance unit is broken.

On cars 1970 and later the procedure may be a bit different. With the advent of exhaust emission controls the newest cars use vacuum advances modified with a device which cuts off the advance at low speed or idle. This reduces the amount of unburned fuel and keeps the atmosphere cleaner.

GM cars of the early seventies use TCS or transmission controlled spark. A solenoid on the transmission allows vacuum to get through in top gear only. However, there is also a temperature sensor built in to the system to allow full vacuum advance when the engine is cold. The advance is needed to prevent stalling.

Chrysler uses a solenoid on the advance mechanism controlled by the throttle valve on the carburetor. When the throttle is closed the vacuum advance is cut off. While the engine is still cold the carb's fast idle mechanism keeps the advance working.

Ford works their system off the speedometer cable. Vacuum advance is cut off at low speeds. There is a temperature sensor to restore vacuum advance in cold weather.

What does all this mean, aside from the fact that—on late-model Fords—the vacuum advance isn't working when the speedometer cable is broken? It means that there may be two vacuum advances on your distributor and two hoses which must be disconnected and pinched off while you time the ignition. Also you might find a little black box (electronic module, they call it) on the vacuum advance with wires running to it. Disconnect these wires along with the vacuum hose when you time the ignition.

REPLACING THE DISTRIBUTOR CAP AND WIRES

Inspect the distributor cap and replace it with a new one if there are small cracks or carbon tracks inside, or if there is a heavy build-up of deposits on the contacts. Otherwise just use a rag and soapy water to clean the cap and then wipe it dry.

When putting on a new distributor cap, take the plug wires off the old cap one at a time and attach them to the new one with both caps held in the same position. This will prevent getting the plug wires mixed up. Many distributor caps have numbers at each terminal indicating which cylinder the wire goes to. In this case all you have to do is match the wires to the corresponding numbers.

Re-install the cap on the distributor by aligning it with the notches or screw holes in the distributor body. It is a good idea to burnish the metal connectors at both ends of each plug lead to ensure good electrical contact. Also, if there is a plastic or wire support (comb) used to keep the plug wires separated, put each wire in its own notch. Crossed plug wires can induce current in each other causing two cylinders to fire at once.

REPLACING THE SPARK PLUGS

Put the new gasket, if there is one, on each spark plug and start it in place by hand. With recessed plugs, use the spark plug socket to start each plug in place, but don't attach the socket to the ratchet handle. You can work more gently by rotating the socket with your hand. If a plug gets cross-threaded it may ruin the threads in the engine block and you will have to pay for having a thread insert installed.

Once the plug is seated well on the threads, you can use your wrench to tighten it. Tighten plugs firmly holding the ratchet handle near its head so you cannot get too much leverage and overtighten the plugs. The manufacturer of your car sets specifications for how much force (torque) to use on the plugs should you happen to have a torque wrench.

Connect the plug leads according to the numbers on your tags.

TIMING THE IGNITION

An inexpensive timing light of the neon variety can work just as well as a better one, but it must be used in a darkened area. Timing lights powered from electric outlets are most reliable. There are also rechargeable ones (expensive) and timing lights powered by leads going to the auto battery terminals.

The timing light wire which gets its impulse from the ignition must be connected to the wire going to the spark plug in the number one cylinder. Sometimes an adapter is necessary to make the connection. The number one cylinder is the one at the front of the engine on four- or six-cylinder cars. On eight-cylinder cars it is the front cylinder on one or the other bank of cylinders. Find out which one by seeing which plug wire terminal the rotor points to when the timing marks are aligned.

General hook-up scheme for ignition test equipment. Tech or dwell meter is connected to coil distributor terminal and ground. Voltmeter is hooked to coil battery terminal and ground. DC-powered timing light draws power from hot battery terminal and is connected to number one cylinder lead wire and ground. (Pontiac)

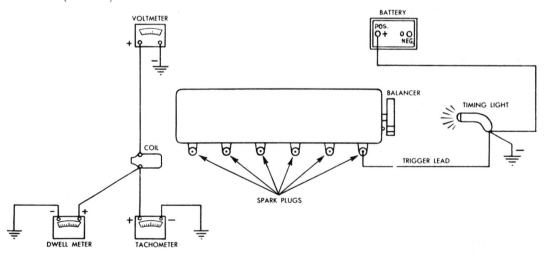

Now locate the timing marks. These are at the front of the engine on the crankshaft pulley or vibration damper, or on the flywheel at the back of the engine. Check your owner's manual or tune-up manual if you can't find the timing marks on your car or don't know which ones to align. Another source of information is *Motor's Auto Repair Manual* which has useful illustrations showing timing mark locations.

To make the marks show up better, wipe all grease off. Then go over both the stationary and rotating marks with chalk or white paint.

Point the light at the timing marks with the engine at idle. The light will be cycling off and on again rapidly, corresponding with the power going to the number one plug. Each time the light flashes on, the timing marks should coincide. If they don't, you will see the rotating mark as a white line separate from the stationary mark.

To make the marks coincide, loosen the distributor lock nut and rotate the body of the distributor. Keep checking until the marks do coincide. Then hold the distributor body down as you tighten the lock nut. Recheck the timing when you are done to make sure that the distributor body hasn't moved.

Late-model cars with emission controls may require timing at a specified r.p.m. Tune-up specifications will be found on a sticker in the engine compartment.

Don't forget to reconnect the vacuum advance line.

STATIC TIMING

Older Volkswagens and many other foreign cars are timed statically with the engine turned off. The static timer is just a light like the one used to troubleshoot the electrical system. It has a 6- or 12-volt bulb (depending upon the voltage of your car) with leads made from 14- or 16-gauge wire soldered to the base. Alligator clips go on the end of each lead wire.

Remove the spark plugs and turn the engine over until the timing marks line up. On a VW you can turn the engine over with a wrench which fits the nub of the generator pulley (it comes in the tool kit supplied with the car and is also the lug wrench).

Attach one lead of the timing device to the distributor primary terminal (that's the terminal which is connected to the coil by a thin wire) and the

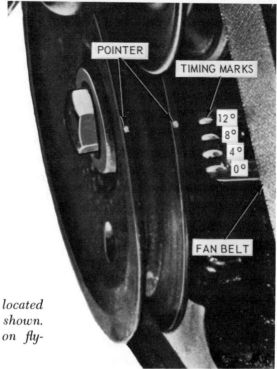

Engine timing marks are usually located on or near crankshaft pulley as shown. Sometimes timing marks are on flywheel at rear of engine. (Ford)

other to a good ground on the engine block or chassis. Then align the timing marks. Turn on the ignition (but don't start the engine) and rotate the distributor body in the direction the distributor cam rotates (after loosening the lock nut) until the light just goes on. Tighten the lock-nut and you are done.

OTHER WAYS

There are any number of devices sold to make tune-ups easier. The best known is probably the tach-dwell meter used by most professional mechanics to set the point gap and idle speed. Instead of measuring the gap directly, a dwell meter measures the dwell or degrees of engine rotation during which the points remain open. Point dwell is another specification given in tune-up manuals.

Rather than getting a tach-dwell meter, which is fairly expensive, a recently introduced device called the Compu-Dwell can be used. It is both a static timing light and a dwell meter which works like a magnetic compass. To use it in measuring dwell, the distributor body is turned until the light on the Compu-Dwell comes on. Then the pointer is zeroed and the distributor is turned further until the light goes off. The dwell angle is then read from the pointer. If it is not according to specification, the point gap is adjusted until dwell is right. Of course, the distributor must be returned to its original position so that timing is not changed. Chalk marks made prior to rotating the distributor will ensure that it is reset at the correct point.

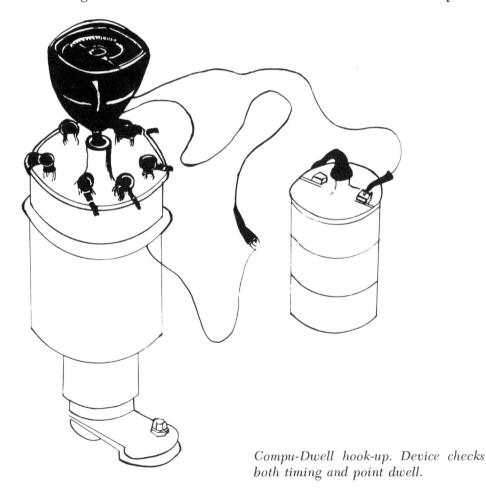

Compu-Dwell hook-up. Device checks both timing and point dwell.

In emergencies there is a way to set the timing fairly closely without using a timing light. To use this method you turn the engine to align the timing marks. Then the distributor lock nut is loosened and the distributor body turned until the rubbing block on the points is just about to contact the cam lobe for the number one cylinder. (You can tell which lobe it is because the rotor will point toward the number one cylinder lead wire.) Then place a thin strip of paper between the points and pull on it with a steady, light pressure. Rotate the distributor body a bit more with your other hand and the points will begin to open, letting go their grip on the paper. Leave the distributor body at just the point where the paper came loose and tighten the lock nut.

ADJUSTING THE VALVES

Most American cars no longer require regular valve lash adjustment since they use hydraulic valve lifters which are self-adjusting. On engines with mechanical lifters the valve lash must be adjusted each time the engine is tuned. Your owner's manual will tell you if this is a required procedure on your car and whether valve lash adjustment is done with the engine cold or at normal operating temperature.

Valve lash adjustment is necessary on Volkswagens and should be done at tune-up time in order to keep the engine healthy.

To understand what valve lash means, remember the explanation of engine operation in Chapter 1. The camshaft is connected to the crankshaft by gears or a chain drive and rotates whenever the crankshaft is turned. Rotation of the camshaft causes the intake and exhaust valves to open and close at the right point in the engine cycle.

The camshaft is not directly in contact with the valves. Instead there are lobes on the cam which move against a valve lifter and operate a push rod going to each valve. The push rod pushes against one side of a rocker arm above each valve and the other side of the rocker opens the valve by lifting it away from the exhaust or intake port on the cylinder. Some engines have a camshaft located above the valves so push rods are not needed.

The valve is closed by its own spring. When the cam lobe operating a particular valve is against the lifter, the valve opens. When the low spot on the camshaft is beneath the lifter, the valve is held closed by its spring.

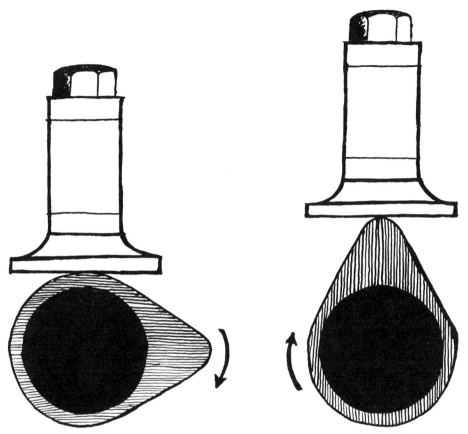

Cam lobe pushing on valve lifter (tappet) opens valve via pushrod and rocker arm. Valve on left is closed; right valve is open.

There must be a clearance maintained between the rocker arm and the valve stem tip to allow for valve stem expansion when the engine is hot. This is the valve "lash."

Clearance specifications for both exhaust and intake valves should be listed in the owner's manual of cars that require periodic valve lash adjustment.

The valves are exposed when you unbolt or unclip the rocker cover. It is at the top of the engine on most cars, but there are valve covers on either side of the flat, opposed cylinder, Volkswagen engine.

Clearance is measured with a flat-bladed feeler gauge inserted between the rocker and the valve stem tip. Sometimes adjustment is specified be-

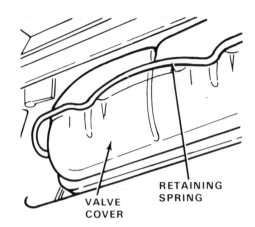

VALVE COVER

RETAINING SPRING

Valve cover on VW can be opened by looping belt around retaining clips and tugging downward. (Volkswagen)

tween the rocker and push rod. To measure the gap correctly, the valve being checked must be closed.

Start with the engine either cold or hot, according to manufacturer's specifications. Rotate the engine manually until the timing marks are lined up. Assuming you have just checked to see that the timing is correct, the distributor rotor will point to the spark plug wire going to the number one cylinder.

Before removing the distributor cap, use chalk or paint to make a mark on the distributor body directly beneath each plug wire attaching point. Then remove the distributor cap and check to see that the rotor is pointing to the number one wire mark.

Unbolt or unclip the valve cover. On Volkswagens you will start with the right side valve cover and you should place a pan under it to catch oil dripping out. The retaining clip is pried downward.

In cases where the valve cover does not come off readily, tap the edges with a wood block. If you must pry, pry gently.

It is useful to have a spare valve cover gasket on hand in case the one in use is deteriorated. If you are replacing the gasket, scrape the old one off and clean the valve cover with solvent, then use gasket sealer on the mating surface of the cover. The gasket is sealed only on the cover side.

Rotate the engine and watch the valves opening and closing to determine which ones belong to each cylinder. Check the firing order for your car in the service manual. Look at the exhaust manifold where it branches off to connect to each cylinder. The location of the manifold branches will tell you where the exhaust valves are. On a VW the intake valves are in the center of each set of cylinders with an exhaust valve at each end. The pair of valves toward the front end of the car belong to the number one cylinder.

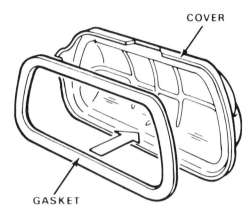

Valve cover gasket should be replaced if worn or broken. (Volkswagen)

Adjusting valve clearance. (Ford)

On Volkswagens, where the firing order is 1-4-3-2, you can rotate the engine backwards (counterclockwise) so the order is 1-2-3-4. For each 180 degrees of crankshaft pulley rotation, corresponding to 90 degrees of distributor rotation, you will be at the firing point for a cylinder and the valves for that cylinder will be closed. If this seems confusing, you can always stick a screwdriver through the spark plug hole to ensure that the piston is at the top of the cylinder. You will remember that when a piston is at or close to TDC, both valves are closed and combustion takes place.

Select the correct feeler gauge and insert it between the valve stem tip and the rocker arm or bottom of the adjusting screw. It should fit snugly, while the next thicker blade should not fit.

If the gap is incorrect, you will have to set it with the adjustment screw. On most cars there is a lock nut to be loosened with a wrench before the screw can be adjusted. A few have self-locking screws. Turn the adjusting screw until a bit of drag can be felt on the feeler gauge blade, then remove the blade, hold the adjusting screw so it will not move, and tighten the lock nut. Recheck to be sure the gap is correct.

When the gap has been set for both valves of number one cylinder, rotate the engine so that the distributor rotor is pointing to number two cylinder and do that set of valves.

Keep turning the rotor and adjusting the valves until you have gone through the complete firing order and all cylinders have had their valves set to specifications. Then replace the rocker cover or covers and put the distributor cap back on.

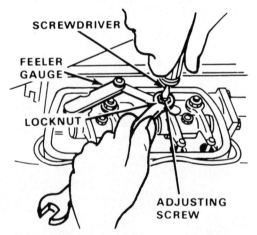

On some cars a lock nut must be held as adjusting screw is turned. (Volkswagen)

ADVANCING OR RETARDING TIMING

Sometimes it is necessary to retard or advance spark timing when modifying an engine or just fine-tuning it for the thinner air of mountainous climes. For instance, a service manual may suggest an advance of three degrees under certain conditions (on pre-emission-era cars). How do you find this point and set a new timing mark?

Moving the timing mark in the direction of rotation on the crank pulley means advancing the spark and moving the mark in the opposite direction will retard the mark.

Now look at the diagram. Angle B is the number of degrees specified, in this case three degrees. The distance C is the radius of the crankshaft pulley, and you can find it by measuring from the mid-point of the pulley to any point on the outer edge. All you need to find point A, your new timing mark, is a little math.

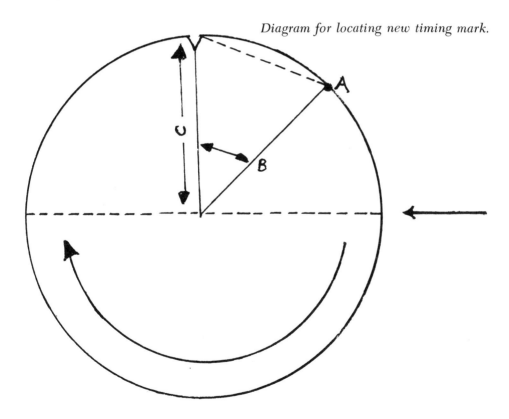

Diagram for locating new timing mark.

For those who dig mathematics, here's the formula:

$$\frac{Distance\ C \times Sin\ Angle\ B}{Sin\ (90\ degrees - \underline{Angle\ B})}$$
$$2$$

Or, for those who hate math, the technique is to find a neighborhood kid who's taking trigonometry in high school and ask him to look up the sine (pronounced "sign") of the angle, in the table at the back of his math book. Multiply the decimal figure he gives you by the distance you measured. Then take half your angle, subtract it from ninety degrees, and ask the kid to look up the sine of that angle. Then divide the figure he gives you into the first figure. Sound complicated? Maybe you can get the kid to handle the whole job. Anyway, when you get the answer, which is the distance from your timing mark to point A, the new mark, all you need is a ruler to locate the new point and mark it.

ADVANCED IGNITION SYSTEMS

The distributor system is the hardest working part of your car. That little set of points, selling for a dollar or two, has the enormous task of making and breaking contact thousands of times per minute at higher engine speeds. It's not surprising that the points may "bounce" at very high engine revs and cause loss of power.

One solution is to have a double set of points and share the work. Dual-point distributors are fitted to a number of performance cars and are available as aftermarket accessories.

There is still the fundamental problem of a heavy primary current (3 or 4 amps) which the points must handle. This is why points are subject to burning and pitting. A transistor, which makes and breaks contact electronically, can reduce this point load.

In one type of transistorized ignition system, the points serve only to trigger a transistor, which handles the heavy current. However, the amperage going through the points must still be high enough to get through greasy contacts and other contamination which may develop.

A straight transistor system would consist of a transistor unit, a distributor minus the condenser, and a special coil with a high primary current.

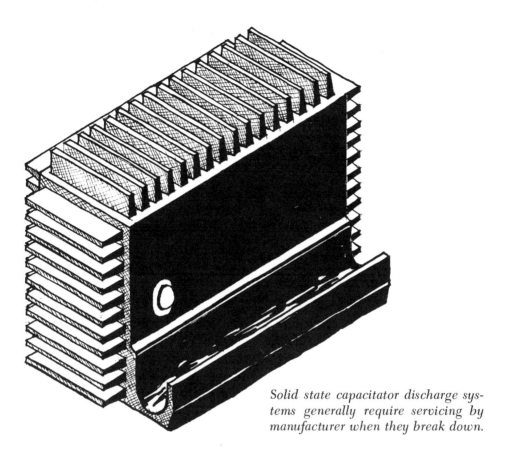

Solid state capacitor discharge systems generally require servicing by manufacturer when they break down.

There are a number of theoretical advantages of such a system, but the only one which seems to be practically significant is longer point life—up to 40,000 miles or so.

Straight transistor systems never became very popular since their cost was high in relation to the savings which could be expected from running the same set of points longer. Also many early versions suffered from reliability problems and had a tendency to cut out suddenly with no warning.

Another version of the transistorized ignition eliminates the points altogether and replaces them with a magnetic impulse generator and an amplifier, or with a photoelectronic (light-beam controlled) system. This requires a special distributor with new innards, plus the transistor unit. Point bounce, burned points, and point gap changes caused by rubbing block wear are eliminated. These systems are costly, but the constant output voltage at higher engine speeds is a boon to performance cars.

At this point we might mention the magneto, which is the ignition system preferred by performance enthusiasts since well before transistor units came along. The magneto can be thought of as a generator, distributor, and coil, all wrapped up in one unit. Light weight, excellent reliability, and precision manufacture are the main advantages of the magneto system.

Although magnetos do use breaker points, they are manufactured to last longer than those on conventional distributors. The popular magnetos, such as the Scintilla Vertex from Switzerland, are precision mechanisms. They are costly, but durable, and designed primarily for racing cars, though adaptable for street use.

The breakthrough for transistorized systems came with the CD or capacitator discharge ignition. A CD unit can be triggered by conventional breaker points in the distributor or by a breakerless distributor. Its advantage is that it stores up current, taking over this function from the coil in conventional systems. In a CD system the coil is used only to step up voltage.

A CD ignition has a fast "rise time," meaning that it develops maximum voltage across the spark gap much more swiftly than conventional systems can. There is no time for this voltage to be drawn off by fouled porcelain or moisture and fail to jump the spark gap. Thus CD systems have a remarkable capacity to fire fouled plugs and they also work better in inclement weather.

A breakerless CD system can keep operating efficiently for long periods of time without servicing. Now that auto manufacturers are under pressure to produce a car that will stay in tune, without emissions increasing, for 50,000 miles or more, CD systems have been adopted on a few American cars and will probably see more widespread use in the future.

There are various means for setting the dwell time on distributors without breaker points. However, servicing the solid state electronic parts of the system is a job beyond most amateur mechanics and some garages. When a solid state system does go—and they usually fail abruptly—it may mean a trip to an authorized repair center. So, before you purchase an aftermarket ignition system, check with the manufacturer about servicing facilities.

7

Starting a Dead Car

● Starting a dead car is not like flogging a dead horse; it requires finesse rather than force. You will want to check for the obvious causes first, then check for a fault in the starting system, ignition system, or gas supply system.

The most obvious cause of all is that you're out of gas. If so, proceed directly to the nearest gas station with an empty can and don't bother to read further.

Try starting again to be sure you went through the right steps. Is the automatic transmission in Park or Neutral position? Did you depress the gas pedal once before trying to start in order to close the automatic choke (or pull out the choke lever on a car with a manual choke)?

Late model manual transmission cars require that you depress the clutch fully to complete an electrical circuit before the starter will function. Have you done this?

If the engine is warm from previous use of the car, keep a manual choke pushed in. With an automatic choke, depress the gas pedal halfway and release it.

Pumping the gas pedal too much or having the choke out on a warm car may result in the carburetor being flooded with a too-rich gasoline/air mixture. Push the accelerator pedal to the floor and keep it there while you try to start. This will open the choke unloader and get more air into the carburetor.

Cars with fuel injection and many British cars with a different type of carburetion will require an altered starting procedure. Follow recommendations in your owner's manual.

If the car won't start on the second try, it's time to check under the hood for the cause of the trouble.

Either the engine cranks over or fails to crank over, or else it cranks over weakly. When the engine completely fails to crank over, or there is an awwr-awwr-awwr sound indicating that the engine is being barely turned over by the starter, the trouble is almost always in the starting system. An engine which turns over vigorously but doesn't "catch" is indicative of trouble in the ignition or gas supply systems. We will cover these conditions separately, beginning with the most common sources of trouble and going on to the rarer ones.

NO CRANKING OR SLOW CRANKING

The most common causes are:

> A dead or weak battery;
> Poor battery connections;
> A starter solenoid which is not making good contact.

Less frequent causes are:

> A starter pinion hung up on the flywheel;
> A faulty starter drive or starter motor;
> A defective neutral safety switch (on automatic transmission cars only).

Start by checking the battery. Do the cells have enough water? Are the terminal clamps tight? Is there sulfating or corrosion around the terminals indicating that electrical contact is poor? Is the top of the battery damp?

Add water to the cells if needed. Take the terminal clamps off and scrape the terminals and the inside of the clamps with a pocketknife or file. Wipe off any moisture on the exterior of the battery. If you cannot get the terminals off easily and clean them, try jamming a screwdriver between each terminal and its clamp and then attempt to start the car. Remember that it takes very little corrosion to destroy good contact and prevent the battery from doing its job.

Trace each battery cable to its other end, where it connects to a ground, junction block, or to the solenoid. Look for signs of looseness or corrosion here too.

A hydrometer can tell you the condition of your battery, but chances are that you don't have one along. Instead, have someone check the headlights while you turn the ignition switch to the start position. If the lights go out or grow dim, chances are that the battery is the culprit and you'll have to try getting a boost.

To boost the battery you will need a friendly neighbor or passer-by who volunteers the use of his car. Jumper cables are used to connect the battery terminals on the two cars.

Have your neighbor draw up his car so that the batteries on the two cars are close enough for jumper cables to reach between them. With the engine off, connect the two negative terminals and then the two positive terminals. Have your neighbor start his car and run the engine at fast idle for a minute or two. Then try to start your car again.

In the absence of jumper cables, you may still be able to transfer battery power from one car to another so long as both cars have the same grounding (positive or negative) for the battery. Draw the cars up so that the bumpers touch. This should make electrical contact between the terminals which are grounded, since they are usually grounded to the frame and body of each car. To get contact between the other two terminals, use your jack or two lug wrenches to span from terminal to terminal.

A similar technique is useful when jumper cables won't stretch from battery to battery. Hook both jumper cables together to stretch between one set of terminals and use bumper to bumper contact to complete the circuit.

Be careful of your hook-up on alternator-equipped cars. Any mismatched polarity—such as might happen if you accidentally brushed the body of one car with a jumper attached to the positive terminal of the other—can put the kibosh on the system.

You can also get started by having another car push yours. All manual transmission cars are push-started in a similar manner. After first looking to see that the bumpers mate, have your car pushed up to about 15 m.p.h. with the ignition on. Place your car in second gear and let out the clutch gradually. The car should lurch a bit and then start.

When a manual transmission car cannot be push-started, it usually means a dead short across the battery, no gas or no spark, or loss of compression in the cylinders.

Remember that cars with alternators cannot be push-started when the battery is completely run down because alternators are not able to build up a charge where none exists.

Auto transmission cars may not be able to be started by pushing due to transmission design. GM Turbo-Hydro transmissions, the Super Turbines, Buick's Dual Path Drive, and Ford's Cruisematic are among the auto transmissions which have no provision for push-starting.

Most other auto transmissions may be push-started by pushing in Neutral with the ignition on. To start, shift to Low at 15 to 20 m.p.h.

The old Dual Coupling Hydra-Matics must be push-started at about 30 m.p.h. by shifting to Drive. Hydra-Matics made after 1958 can't be push-started at all.

Ford two- and three-speed automatic transmissions must be shifted to Low at about 25 m.p.h., but three-speed versions made after 1967 can't be push-started.

Solenoids and Starters

When the battery has been checked out and found to be functioning adequately, look to the solenoid next for your source of trouble.

Since the solenoid is a type of switch, it can be jumped or cut out of the circuit. On Ford products, where the solenoid is sitting right out in the open on the sidewall or firewall, you can bridge the solenoid by running a jumper cable from the big (thick-wire) terminal on one side of the solenoid to the similar terminal on the other side. You can even do the job with a screwdriver blade.

On other cars it may be more convenient to run a jumper from the positive battery terminal to the main starter terminal. However, on Volkswagens and other rear engine cars where you have to crawl under the car to reach the starter situated in front of the engine, it is easiest to short across the two large solenoid terminals.

Assuming that your car still stubbornly refuses to function, and that you have an automatic transmission, the trouble may be in the neutral safety switch used on auto transmissions to make the starter function only when the selector lever is in Neutral or Park positions. If you jumped directly from the battery to the starter when bridging the solenoid, you also cut the ignition switch and neutral safety switch out of the circuit. If not, you will have to bridge these also to be sure that they are not faulty.

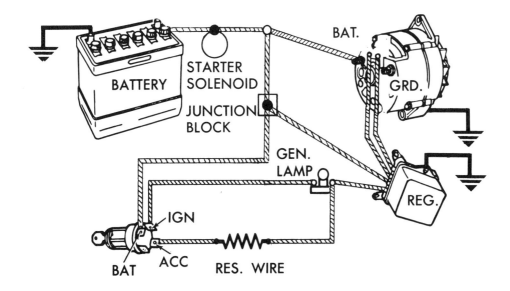

Starting and charging circuits. This is a general schematic for GM cars. (Chevrolet)

KEY TO ABBREVIATIONS: *ACC.—accessory terminal; BAT.—battery terminal; GEN. LAMP—generator lamp; GRD.—ground; IGN.—ignition terminal; REG.—voltage regulator; RES. WIRE—resistor wire*

But first try jiggling the automatic transmission selector lever in both Neutral and Park positions while you attempt to start the car. Sometimes this will cause a loose neutral safety switch to function.

To bridge the neutral safety switch and cause the solenoid to function directly, run your jumper from the hot terminal of the battery to the small terminal on the solenoid.

Still no luck? Then move on to the starter. Some procedures for troubleshooting the starter were covered in Chapter 4, so we'll just briefly review them here.

Sometimes the starter drive gear gets hung up on the ring gear of the flywheel. On manual transmission cars you can put the car in a forward gear and rock it backward to free the starter drive. With an automatic transmission all you can do is bang on the starter case and hope this will free the pinion. Or you can take out the starter and re-install it again provided that it's relatively accessible.

Starter brushes which are making poor contact can often be temporarily fixed by wedging the brush springs with a bit of cardboard.

At this point, if you still aren't getting a response, you may be about

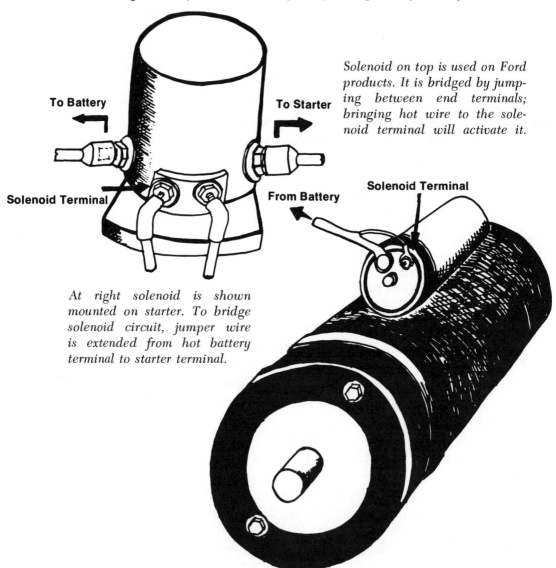

To Battery

To Starter

Solenoid on top is used on Ford products. It is bridged by jumping between end terminals; bringing hot wire to the solenoid terminal will activate it.

Solenoid Terminal

Solenoid Terminal

From Battery

At right solenoid is shown mounted on starter. To bridge solenoid circuit, jumper wire is extended from hot battery terminal to starter terminal.

ready to give up. However there are a few special conditions to check for before you begin hiking to the nearest service station.

On a very cold day the engine may crank over slowly but never start because the oil is too heavy. Unless you have a heater or campstove around to warm things up, getting a push-start is your best solution. When you get home, drain the oil and put in some lighter weight stuff.

When you can't get any response at all from the engine, even after trying your best, it's a wise idea to pull one of the spark plugs to check for fluid in the cylinders. A bad head gasket could have caused water to leak into the cylinders, or perhaps the carburetor float may have gotten stuck and caused the cylinders to fill up with fuel. Pulling all spark plugs to drain the cylinders will get you started again and on your way to the nearest gas station where the trouble can be checked out.

Perhaps your engine is turning over, but it cranks very slowly. You naturally assumed that the trouble was in your starting system. However, it could be an ignition problem. So let's move on to checking out the ignition and fuel systems.

THE ENGINE CRANKS BUT WON'T START

When the engine does crank over, especially when it cranks over vigorously, assume that the problem is right at the cylinders where combustion is taking place. Either no gas/air mixture is going to the cylinder, the spark plug is not sparking (or not creating a strong enough spark), or else the piston is not moving down in the cylinder due to a mechanical failure in the engine. Fortunately, the latter condition rarely happens without plenty of forewarning.

Testing for spark.

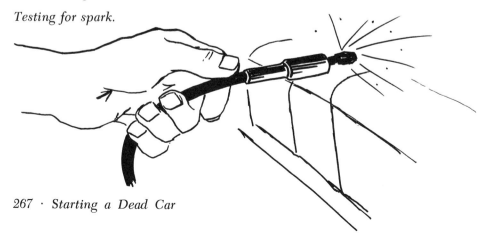

267 · *Starting a Dead Car*

The first step is to check for spark and then for gas. Check for spark by pulling one of the spark plug wires off the end of a plug and pulling back the rubber boot until the metal connector is exposed. Grasping the wire by its rubber part only, hold the metal connector one-eighth of an inch away from a ground such as the engine block. Then have someone crank the engine. No spark or a weak, yellowish spark means the trouble is in the ignition system somewhere and you will have to trace it.

In case you cannot pull the rubber boot on the spark plug wire back to expose metal, use a key to contact the metal and extend outward. Keep the key one-eighth of an inch away from the engine block while someone cranks the engine.

Testing for gas by hand operating accelerator linkage. Further test may necessitate removing gas line to carb.

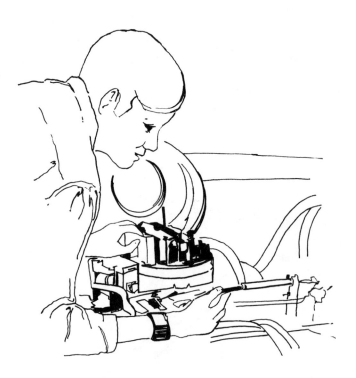

No one to crank the engine while you make the test? Then do it yourself by bridging the solenoid and test for spark as the engine cranks over.

A healthy spark means the trouble is probably in the fuel delivery system. You cannot check for this at the cylinder without unbolting the intake manifold. However, there is not much in the intake manifold to go wrong, so check for gas going into the carburetor.

Take off the air filter canister and push open the choke plate so you can look into the carburetor. As you peer into the carb, use one hand to pull on the accelerator linkage at the side of the carburetor. This is the mechanical linkage from the gas pedal and it should squirt a jet of gas into the carb each time it is pulled. If you can't see the gas you can smell it and hear it squirt.

The Next Step

By now you have either traced the problem to a failure in the ignition (spark-making) system or the fuel delivery system. In rare cases where there is gas and adequate spark, the diagnosis is a mechanical problem in the engine. To confirm this diagnosis take off the distributor cap (leave the plug wires connected) and see if the rotor goes round and round as you crank the engine. The rotor is driven off the camshaft. Its failure to rotate probably means that the camshaft is broken or the timing chains between the crankshaft and camshaft have snapped. The only thing to do is call a tow truck.

Tracing the Spark

You have traced the problem to a weak spark or no spark. Now you have to find out which part of the ignition system is falling down on the job.

First grasp the body of the distributor and try to rotate it by hand, using considerable force. Also try to move it up and down.

A distributor which moves will throw the timing way out, so there's your problem. To correct it, try rotating the distributor both ways a little as someone cranks the engine. When you get the distributor to a point where the car will start, drive to the nearest garage and have the timing checked and the distributor lock nut tightened. (Or do your own tune-up following instructions in Chapter 6.)

If the distributor is tight, check the secondary (high-voltage) ignition circuit first. You have learned that this is the circuit that goes from the coil (which creates the high voltage) to the top of the distributor and then to the spark plugs.

Check for spark coming out of the coil by pulling off the thick wire which goes to the middle of the distributor. This wire comes from the coil and carries high voltage. Make the spark test the same way you did before; place the metal connector on the wire end about an eighth of an inch away from the engine block and crank the engine. No spark or a weak spark means the coil may have to be replaced. The other alternative is that not enough low-voltage current was coming to the coil and requires a check of the low-voltage (primary) circuit. We'll do that after we finish with the high-voltage circuit.

Now let's say there was a strong spark at the coil wire and a weak spark or no spark at the plug wire. Obviously the trouble is somewhere in the top of the distributor where the high voltage is sent to the plugs.

Pull off the distributor cap and check the inside for small cracks or carbon deposits. Either of these conditions can divert voltage from its correct path to the spark plugs.

Scrape away carbon deposits with a pocketknife. Use the point to clean the metal contacts which the plug wires fit. Remove only one plug wire at a time so you don't forget where each one attaches.

Cracks in the distributor cap can be filled in with gasket sealer, but there may be an easier way. It is not the crack itself which diverts the voltage but moisture which collects in the cracks. Punch a small hole through the cap at the mid-point of the crack and you should be able to get going until you can obtain a new distributor cap.

Clean the metal tip of the distributor rotor if there are any burn marks on it. Sand or file the tip gently and bend it upward slightly to ensure good contact.

Wipe any moisture from the spark plug wires before putting the rotor and distributor cap back and trying to start the engine again.

In cases where there is a strong spark from the coil and the engine still fails to start after you have replaced the condenser and cleaned the distributor cap and rotor, the fault is in the spark plug wires or the plugs themselves. Check the spark at each plug to see which wires are defective. You can replace one temporarily with a heavy jumper wire.

Pull the spark plugs one at a time with a plug wrench and wipe any traces of grease from the plug body and electrodes. Inspect the ceramic part of the plugs for cracks. Scrape off any burnt deposits with a pocketknife. If you don't have a feeler gauge to check the plug gaps, use two thicknesses of a matchbook cover which is approximately the right gap for all newer cars. On a very old car use a single thickness of matchbook cover. Carefully bend the electrode to the right gap. Replace each of the plugs and the engine should start

Low-Voltage Circuit

Now let's go back to the point where we checked for spark at the coil wire. If there is no spark or the spark is weak it means a faulty coil or not enough low voltage being supplied to the coil. The third possibility is that the coil wire is defective. This can be checked by temporarily disconnecting a spark plug wire and using it to substitute for the coil wire as you make the engine-cranking test for spark. Should there be a healthy spark it means the coil wire is bad and needs replacing.

The next step is to check the low-voltage circuit. The two thin-wire coil terminals on either side of the thick high-voltage circuit terminal are both connected to the battery (low-voltage) circuit. One wire indirectly runs from the positive (ungrounded) terminal via the ignition switch. The other is in circuit with the battery by running through the points in the distributor and thereby being grounded whenever the points close. When the points close they complete the circuit, cause an electromagnetic field to be built up in the coil, and boost the low voltage to high voltage.

To check for low voltage going to the points, flick open the points with a knife blade or screwdriver, with the ignition turned on. There should be a healthy spark as the points open. This would indicate that low voltage is coming to the points and the points themselves are faulty.

The points must be at the correct gap when open. Check the point gap and adjust it following the procedure in Chapter 6. Since you probably won't have a feeler gauge handy, see that the points are apart a distance equal to the thickness of one matchbook cover when fully open.

Check the points for burning or pitting as recommended in Chapter 6. You probably won't have a FlexStone to file the points correctly, so use a nail file or even an abrasive stone. The points won't last long this way, but the car will start if pitted or burned points are the problem.

Now let's see what to do if there is no spark when the points are flicked open. This means no low voltage is going to the points. Perhaps the cause is only a defect in the thin wire running between the negative terminal of the coil and the points. You can check the low voltage at the coil by connecting a jumper wire to this coil terminal and holding it a fraction of an inch away from the engine block with the ignition on. If there is spark it means that the thin wire between the coil and distributor points needs to be replaced.

Now test the coil itself. This will require that you take a small light bulb out of some accessible light in the car to use as your tester. Try a dome light, a dash light, or even a tail light. Don't use a flashlight bulb as it is not the correct voltage.

Block the points open and hold the side of the bulb base against the engine block to ground it. Be sure the ignition is on. Then disconnect the wire from the negative terminal of the coil (the one which connects to the distributor). Touch the wire to the base of the bulb while keeping the bulb grounded. The bulb should light, meaning that the coil is getting juice.

Complete the test by reconnecting the wire from the negative terminal and disconnecting the other thin wire coming from the coil's positive terminal. Touch this to the base of the bulb while it is grounded. The bulb should *not* light.

If the bulb lights when touched by either wire, the coil is defective.

Assuming that the coil checks out, and the battery is putting out enough power to crank the engine, the problem must be in the ignition switch or wiring.

On most cars you can jump (or bypass) the ignition and wiring by connecting your jumper wire from the positive (ungrounded) terminal of the battery to the positive terminal of the coil. (A few late model cars are exceptions.) Remember that the positive coil terminal is the one that does *not* connect the thin wire from the distributor. The positive coil terminal is usually marked + or S or SW or BATT.

Connect the jumper wire with the ignition on. If the car starts, leave the wire in place while you troubleshoot the wiring and ignition switch or have a mechanic check them out. Don't leave the jumper in place too long because you are bypassing a resistor which controls the amount of voltage going to the coil. Without the resistor in circuit the coil won't last long.

Other Ignition Problems

You have made a detailed check of the ignition system and pinpointed the problem causing lack of spark. The only thing you haven't checked is the timing. The spark may be getting to the plugs all right, but if it arrives when the cylinders aren't full of a compressed gas/air mix ready to be combusted it won't do any good.

Ways to time the engine are given in Chapter 6. Use the method detailed at the end of the chapter if you have no timing light along.

Another possible problem is a faulty resistor. Resistors are used to limit the voltage going to the coil. Resistance wiring is used on most cars, but Chrysler Corporation uses a ceramic resistor device. When a car starts but keeps on stopping immediately, the problem is caused by a faulty resistor. Connecting the hot terminal of the battery to the positive terminal of the coil will bypass the resistor as well as the ignition switch and enable you to get down to the repair shop.

Finally, a weak spark may be caused by moisture in the ignition system. On a damp or rainy day try wiping off the coil, the distributor (check inside for moisture), the exposed part of each spark plug, and all the ignition wiring, including the terminal connectors. You would be surprised how many times moisture is the culprit.

Tracing the Gas

Just as you checked back through the ignition system to find out where spark-producing voltage was lost, you will have to check back through the fuel supply system to find out why no gas is coming into the carburetor.

Before you check for gas, however, check for air. Remember that the engine burns an air/gas mixture and lack of air can be just as big a problem as lack of gas.

Perhaps the trouble is only that the air filter is thoroughly caked up with dirt. Try starting the car with the filter off and see what happens.

With the filter off you can see whether the automatic choke is working. On a cold engine the choke plate at the top of the carb should be closed so that very little air gets in. A cold engine needs a rich mixture (much more gas than air) to start properly.

If the choke is open and the engine is cold, put the palm of your hand over the mouth of the carb to choke it manually, while someone cranks the engine. A warm engine and a closed choke plate present the opposite problem. Keep the choke plate open with your fingers while someone cranks the engine. If the choke plate is not opening or closing when it should, see Chapter 3 for details about adjusting the mechanism.

The carburetor itself may be flooding with gas. Most frequently this is caused by dirt which gets into the needle valve and holds it open. Even with an in-line fuel filter this sometimes happens. If so, there will be a strong smell of gas under the hood.

You don't want to poke around inside the carburetor because it's a complicated mechanism. However, there are three ways you can try to dislodge any foreign matter. One is to simply tap the carb around the fuel inlet. Give it a few sharp raps with a screwdriver handle. Another way is to disconnect the fuel line at the carburetor (see below) and pinch it shut to prevent gas coming out as someone cranks the engine. The engine will probably start and use up the gas already in the carburetor. Dirt will be dislodged as the carburetor works.

A third trick is to reverse the leads on two spark plugs opposite each other in the cylinder firing order. (Check the firing order for your car in your owner's manual.) This will cause backfire and should blow out the dirt in the carb.

Try opening the choke plate on top of the carb and pouring a few ounces of gas down in it. Do this with the ignition off. Then start the engine. It may start and continue running even with a gummed up carburetor or faulty fuel pump once it is initially primed.

If the engine does run but stops when the gas is used up, disconnect the fuel line at the carb to check the rate of fuel delivery. The fuel line may be steel or made of a rubber-like material. Soft hose is generally held on by hose clamps. Steel fuel lines are usually connected by a fitting which threads into the carb and another fitting which threads into the first. Two wrenches (one to hold and one to turn) are required to unfasten this type of connection.

Hold a container under the open end of the fuel line while someone cranks the engine. (Avoid smoking while you are doing this and hold the container away from the engine where a stray spark can't reach it.) The gas should come out in regular strong spurts. If it doesn't, trace the fuel line back to the in-line filter and unclamp the filter from the line. Make the same test for fuel delivery at the end where the line feeds into the filter. Gas here and no gas up above means that the filter was clogged. In emergencies you can punch through the filter and replace it in the line until you get to a service station. Just remember that all that unfiltered gas is going to your delicate carburetor.

Slow gas delivery at the filter is almost always caused by a defective fuel pump. See Chapter 3 for instructions on replacing the fuel pump.

Hot Gas and Cold Gas

When a car has been run for a while on a hot day and then refuses to restart, the condition may be caused by hot fuel vaporizing before it reaches the carburetor. This is called "vapor lock" and the only thing you can do about it is wait for things to cool down. You can hasten the process by soaking rags in cold water and wrapping them around the fuel pump and fuel line. If the condition is chronic, the pump and line may have to be shielded or relocated to a cooler position.

The opposite condition from vapor lock is when the gasoline gets too cold and the carburetor or fuel line ices up. The gas itself does not turn to ice. It is water from condensation in the fuel tank which causes the problem. The remedy is to buy a few cans of gas line anti-freeze (dry gas) and pour one in the gas tank every second time you fill up.

SUMMARY

We have not covered all possible causes of failure to start because the serious internal engine problems, starter difficulties, and alternator deficiencies cannot be fixed while you're trying to get to work in the morning or leaning over the hood in some strange parking lot. At least 95 per cent of the starting difficulties you are likely to encounter can be found using the procedures detailed here and repaired at least temporarily.

Although these techniques for starting a balky engine may seem complex and time-consuming, in most cases you will quickly isolate the problem to the starting, ignition, or fuel supply system and will only have to go through a part of the procedure. Often you will have a hunch about what's wrong and be able to skip a few steps.

Always check the obvious areas first. If the engine won't crank over, chances are much better than even that the battery or battery connections are the root of your dilemma. When the engine does crank over, and you seem to be getting both gas and spark, suspect a flooded carburetor. And on wet days, wipe moisture off the plugs and externals of the ignition system before looking further.

8

Trouble on the Road

● "Mexicaneering" is the term entrants in the famed Baja California off-road race have given to repairs made by contestants who use every conceivable field expedient to keep their entries rolling.

These are the guys who substitute pieces of cactus for broken coil springs, use a boulder to beat a bent tie rod straight, and weld parts together using jumper cables from two batteries to create an arc.

While you will probably not encounter the more esoteric car troubles brought on by racing 1000 miles over roads which are almost non-existent, the Mexicaneering kind of ingenuity is still a major factor in meeting roadside emergencies. No sophisticated tools are necessary; just your wrenches and other common hand tools plus wire, tape, nuts and bolts and miscellaneous fasteners of all types, spare parts, cans of gas and oil and other fluids, and a good flashlight. The components of a comprehensive emergency road kit can be selected from tools and spare parts listed in Appendix B.

Roadside repairs are often crude, makeshift, and temporary in nature. The object is to get going; to nurse your car home or to the nearest service station without causing irreparable damage to any major components.

The first sign of trouble on the road may be an instrument reading which is abnormally high or low, a strange noise from under the hood or elsewhere, or simply an engine that stalls out. If the car will run, you must decide whether it is safe to continue or whether you should stop immediately and try to find the cause of the trouble.

MAKING THE DECISION

If your water temperature gauge is creeping toward the danger zone, try speeding up as much as you can with safety. More air will get to the radiator, and fluid in the cooling system will circulate faster. Even if you are stuck in stop-and-go traffic, turning the engine over faster with the car in Neutral will lead to better cooling due to more rapid fluid circulation. Turning on the heater will also help by dissipating some of the heat in the circulating radiator fluid.

Once your temperature gauge needle hits the danger zone, or your cooling system warning light comes on, you should stop immediately.

When the ammeter shows that the electrical system is discharging rather than charging, or the warning light comes on, turn off any power-consuming accessories you can and try to make it to the next gas station. You should stop immediately if your oil pressure gauge reads critically low or the warning light flashes on. Otherwise you might damage your engine.

Unaccustomed noises in your car's operation are another tip-off that you may have to stop right away. For instance, a slapping noise which seems to originate from one of the tires may indicate that a part of the tread has pulled away and is knocking against the fender well each time the tire revolves.

A good rule of the thumb is that the louder the noise, the more serious the trouble. The clatter from a worn connecting rod bearing is quite different from the sound when a connecting rod lets go and all hell breaks loose under the hood.

Engine noises require some experience to evaluate. A crankshaft knock, piston slap, valve click, or the metallic rattle from a loose piston pin are all distinctive, but other noises are more difficult to pin down.

As long as the engine is not overheating and oil pressure is all right, a strange sound under the hood is not sufficient reason to stop and call a tow truck. You should check the oil level and continue, at a reduced speed, until you reach a service station.

Noises which seem to come from the wheels should always be investigated. The problem may be only loose lug nuts, or the whole wheel may be ready to come off due to a loose bearing retainer nut or excessive axle end-play. A thump-thumping sound from the rear may be from a dragging exhaust pipe or a shock absorber which has come loose at one end.

Other distinctive noises are rear axle whine (usually caused by the ring and pinion gear being too tight or lacking lubricant), grinding sounds from wheel bearings running dry, U-joint clunk (too much play in the U-joint), alternator squeal (faulty bearings), power steering squawk (bad valves in the pump), automatic transmission whistle or hum (various causes), and the strident buzz from a speedometer cable lacking lubrication.

None of these noises is necessarily cause for stopping immediately. Since you cannot do any major engine, chassis, or transmission work along the roadside, it is best to proceed with caution and stop only when a noise seems to be getting progressively louder or when the car starts to perform so erratically that it is unsafe to continue on.

Although diagnosis of specific problems may require an educated ear, it is not hard to sense "expensive" sounds and stop before further damage is done.

Naturally you cannot ignore failing brakes, wheel shimmy which is so bad that you cannot control the car, or steering which seems excessively loose or hard. Stop immediately.

COMING TO A STOP

Even if your front hood flies up and you can see nothing through the windshield, try to remain calm and look at the sides of the road for a place you can pull off as far as possible.

When you come to a stop, set the brake and turn the flashers on while you go out to place some flares. If you have an old model car without flashers, leave a signal light on.

You can set out either flares or the red "cat's-eye" reflectors on stands. A set generally comes with red flags for daytime use.

Flares show up best, but they only last for fifteen or twenty minutes. To light them you tear off a cardboard cap and strike the surface under it like a kitchen match. A flare which won't light by striking can be lit by applying a match or cigarette lighter to the head. Do this outside your car.

Use at least two flares or reflectors if you are close to the road or on it. Place one a few feet behind the car and the other at least 200 feet behind it. Another reflector between the first two is a good precaution.

Keep coveralls or an old shirt in the car to put on when you work. You can use a rag, newspaper, or folded-out road map to kneel on. Now you are ready to begin troubleshooting.

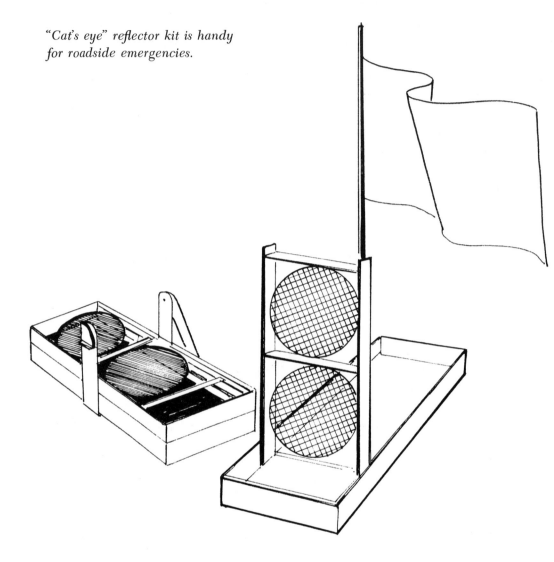

*"Cat's eye" reflector kit is handy
for roadside emergencies.*

STALLING OUT

Procedures for starting a dead car were covered in Chapter 7. You will want to check for spark at the plugs and gas going to the carburetor, then make a detailed check of the defective system.

Maybe you simply ran out of gas? Let's hope you have a spare can along. One gallon of gas won't take you very far so you should be carrying a two-and-a-half-gallon can.

Pour most of your gas in the tank, then take off the air cleaner and pour a little bit into the throat of the carb. That way you won't have to crank the engine until the gas pump can pump gasoline up.

When stopped in the dark, you need some light to do any work under the hood. In my experience flashlights which are kept in a tool kit often don't work when you need them. Leaky batteries are the most common cause. A flashlight with an industrial case and shock housing around the bulb seems to hold up best. Alkaline batteries last longer and don't leak as badly. Substituting a PR-6 bulb for the standard PR-2 will make batteries last much longer although the bulb is not as bright.

You can also get a 12-volt light with leads which clip on to the battery terminals. In an emergency you can take a headlight out and use jumper wires to connect its terminals to the battery.

Jumper cables are an essential part of any emergency kit. When you are working under the hood with no one to crank the engine while you test for spark or gas, a jumper cable to activate the starter is essential.

OVERHEATING

Steam coming from under the hood, plus a water temperature gauge in the danger zone, means that you had better stop right away.

When you come to a stop, do not shut off the engine immediately. Let it idle while you make a visual inspection of the radiator system looking for leaks. After the radiator cools down you may be able to pinch off a tube above a radiator leak with pliers. A can of sealant added to the coolant will also work if the leak is small.

Other substances such as oatmeal are reputed to be effective radiator sealants. However, if you do not carry a can of stop-leak, you are not likely to have oatmeal along.

If the leak is a result of a break in a radiator or heater hose, there is a good field expedient repair. After waiting for things to cool down, wrap the hose in plastic tape or duct tape, extending the repair a few inches above and below the split. Then cover the tape with self-hardening Permatex and let it harden for about thirty minutes. You can finish off the repair with another layer of tape.

It is possible that a hose clamp is loose or missing. Use wire to hold the hose on, after taping around the hose so the wire will not cut into it.

Other parts of the system to check for leaks are the heater hoses, the small bypass hose which runs between the water pump and the engine (if there is such a hose on your car), and the tubes which carry coolant to the automatic transmission. (However, on most cars, the transmission fluid is carried to a tank on the radiator to be cooled.) The water pump itself could be leaking also. Use Permatex to make a temporary repair.

When there are no external leaks, it is possible that water is leaking into the engine due to a defective head gasket. Check the oil to see if the level is overfull. Another way to determine if there is water in the oil is to splatter a few drops from the dipstick on a hot engine. If it sizzles, there's water in it. Be very cautious about driving any farther.

Water may have been coming out of the radiator overflow hose when you checked while the engine was still running. This means that the water level was fairly full and the cause may not be a leak. Look for a missing or broken fan belt, defective radiator cap, faulty thermostat, plugged-up radiator, or a water pump which is not working.

See if the radiator cap was left loose by the last gas station attendant who checked the coolant level. Since modern cars have pressure systems which allow the water to get well over 212° F. before boiling, a loose cap can cause overheating by not maintaining the pressure.

The pressure valve in the cap may have a weak spring. Since it is hard to spot a bad pressure cap without test equipment, it is a good idea to carry a spare and replace the cap when you suspect problems.

Quite frequently a loose or missing fan belt is the cause of overheating. If it is just loose you can tighten it by prying on the alternator or generator after loosening the bracket bolt. (See Chapter 1 for procedures.) Otherwise you will need a spare fan belt which fits your particular car. On a very temporary basis a belt can be fashioned from a rubber shock cord or even a nylon stocking.

Another cause of overheating is a faulty thermostat which is sticking in the closed position. Thermostats are designed to aid engine warm-up by preventing water from circulating through the radiator until the water in the engine passages is warmed up to about 180° to 200°. A thermostat which is stuck in the closed position can cause overheating fast.

The thermostat is at one end or the other of the upper radiator hose.

Loosen the bolts of the thermostat housing and pull it out. Scrape away loose gasket material so it does not get into the radiator. You can run the car with the thermostat out until you locate a replacement.

It is probably necessary to add coolant to the radiator after it has overheated for any reason. When the radiator has been boiling, wait at least fifteen minutes until you attempt to take the cap off. Then turn your face away from the radiator and slowly unscrew the cap to the first detent. Let all the steam come gushing out before you attempt to take the cap off. Use caution to avoid getting burned.

It is a very good idea to keep a jerry can full of water and anti-freeze mixture in your car. Otherwise you must obtain water from some nearby source. Even water from a clean stream will do in an emergency (it can be carried in a hub cap). Add water slowly with the engine running to avoid cracking the block. When you get to the next gas station, drain the water out and replace it with fresh coolant. Have the radiator flushed first, if necessary.

A car which chronically overheats for no apparent reason may have a plugged radiator. The only solution is to have the radiator core removed and cleaned. It is not a job you can do by the roadside.

Sometimes the cooling system is just not up to the job. A heavy load, or pulling a trailer with a lot of hills to climb on a hot day, may be too much for your cooling system. All you can do about the overheating is stop and wait for things to cool down, then add coolant. See about getting a larger radiator or a fan with more blades.

NO POWER AT THE WHEELS

When the engine starts but the car won't go forward, the problem is usually in the transmission or drive train. One exception is when the engine seems to work all right, but stepping down on the gas pedal has no effect on speeding up the engine. The accelerator linkage has come loose.

Look under the hood and trace the linkage from the carburetor. You can fix the loose or broken linkage member by wiring two pieces together tightly or replacing a loose nut.

More difficult to repair is a clutch which fails to engage or disengage. Clutches usually start to slip and then fail gradually. When a clutch goes suddenly, suspect a broken linkage rather than a failure inside the bell-housing.

If you can crawl under the car without elevating it on a shaky jack, or you have jackstands along, try to trace the clutch linkage from the pedal. Have your passenger (or a helpful bystander) push the clutch pedal in and out so you can spot the linkage movement. See where linkage rods are broken or detached and wire or bolt things back together.

On cars with a hydraulic clutch linkage, leaking fluid or air in the system is a likely cause of failure to operate. Have someone pump the clutch pedal while you look for the leak at the master cylinder (usually on the firewall), slave cylinder (on the side of the bell-housing), or the fluid line between the two.

When the clutch is not functioning, it is always possible (if a little foolhardy) to try shifting without using the clutch. The technique is to pause in Neutral and slip the shift lever into the next gear with the engine revved up to just the right speed. Don't try it without prior experience and a tachometer on your car or you could easily wreck the transmission.

The transmission itself may be jammed into gear and the shift lever won't budge. Try tugging and yanking at the shift lever or even hitting it with a rubber-faced hammer. If this doesn't work, manipulate the linkage rods on the transmission case. It helps if you have had prior experience as a chiropractor.

Should you fail to get a transmission out of gear, it is often possible to keep going using that gear only. For high gear you will need to slip the clutch in gradually or start on a downhill slope. If the shift is stuck in reverse gear you may develop a crick in your neck before you get home.

Automatic transmissions which fail to function are not repairable on the road unless the cause is an exterior leak which you can patch and then add fluid. Driveshaft, U-joint, and axle problems will probably require professional fixing unless you're the Baja racer type who manages to wire or weld broken parts by the side of the road and keeps on truckin'.

LEAKS

Small oil leaks can be temporarily checked with soap or a commercial product called Oyltite-Stik. Big leaks due to loss of a drain plug in the oil

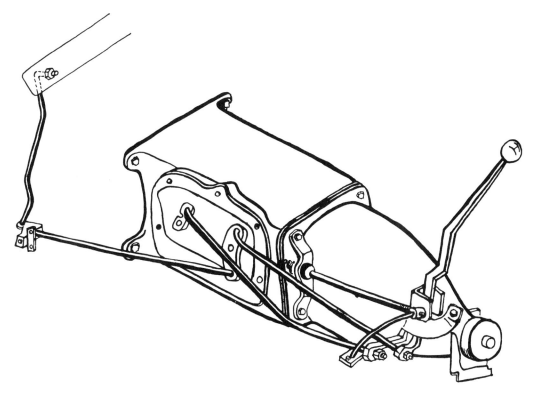

Shift linkage on a floor shift.

pan, gas tank (if it has a drain), or auto transmission pan may be fixed by using a soft rubber universal drain plug replacement sold in auto accessory stores. Large leaks which cannot be plugged will require patching.

You can't successfully patch a gas tank leak with tape, chewing gum, or any other material soluble in gasoline. Instead, try working soap or a screw into the tank to hold the leak temporarily. Then cover it with a few layers of self-hardening Permatex.

With fiberglass cloth and epoxy you can make a more permanent repair, but it will take six hours or so for the epoxy to dry. Instructions for repairing a gas tank leak with epoxy are given on page 50.

In a real, real emergency you can use substitute fluids to refill the braking system or automatic transmission reservoir—even the engine oil can be supplemented with a little kerosene or another solvent to temporarily raise the oil pressure so you can get home. Anti-freeze (or water and anti-freeze from a radiator mix) will do for brake lines and for an auto transmission or power steering reservoir. Do not use gasoline to mix with anything as it is highly flammable. Drain any substitute fluids promptly when you reach the first service station. Flush the system before filling with the correct fluid.

BRAKES

Brakes rarely fail completely these days because modern cars have a dual hydraulic system with the front and rear wheel brakes on separate lines and two compartments in the master cylinder.

When they do give out altogether, look for the problem at the master cylinder. It is possible that the entire unit has come loose and detached itself from the brake lines, or that a negligent mechanic left the cover off the cylinder.

When the cylinder has come adrift, you may be able to wire it back into position and replace the fluid. Tighten the lines to the cylinder well since there is a fair amount of hydraulic pressure at the connections. If you have any doubt at all about whether your repair will hold, it is best to proceed with extreme caution. Driving without brakes is nobody's idea of fun.

On an older car with a single hydraulic system for all brakes, a leak in one brake line or wheel cylinder can drain fluid from the whole system and cause brakes to fail completely. Have someone pump the brakes while you inspect the brake lines to find the leaky point. A leak in the wheel cylinder can be spotted by looking at the back of the wheels. Also you can often smell a brake fluid leak since the stuff has a pungent order.

When you find a leak in a wheel cylinder, remove the brake line from the backing plate and pound the end of the steel line closed with a hammer or rock. Double the line back on itself and pound it flat again to make sure it holds. Add more fluid and bleed the brakes, checking for leaks again.

A leak in a brake line may be repaired by tightening connections and putting Permatex around them, provided that you use this only as an emergency procedure and drive ultra-carefully until you can get home and make a better repair. If there is a break in the line you may have to cut the line off with a hacksaw upstream of the leak and then pound it flat as above.

Using the technique of closing off a brake line to contain a leak in the line or wheel cylinder will give you three-wheel brakes. Although they will be uneven in application and should be used cautiously, chances are you will be able to get to the nearest service area where permanent repairs can be made.

ELECTRICAL SYSTEM

Running through a puddle and splashing up water can drown out your ignition system. Turn off the ignition key and see if you can push the car to a dry place to work. Then dry the spark plug and coil wires and all components of the distributor. Dry all terminal connectors also.

Lights, wiring, and switches are all common causes of roadside troubles. Troubleshooting the electrical system is covered in Chapter 4.

Headlights usually have a circuit breaker in the switch, but most of the other electrical components, including the ignition system, are fused. When you replace a fuse and the new one blows, look for a short circuit in the wiring or switch. As a temporary measure, when you don't have a replacement fuse on hand, you can connect a jumper wire between fuse terminals in the box or even use a paper clip to make the circuit connection broken by a blown fuse. *Caution:* a fuse is a safety measure to prevent overloaded circuits and electrical fires—running without a fuse in a defective circuit can be dangerous. Proceed with caution and stop from time to time to check for smoldering wires.

When an ignition switch won't function (or you lose the key), you will have to jump the switch. On older cars you could do this by connecting a jumper wire between terminals on back of the ignition switch. Newer cars have the terminals inside the switch to foil would-be car thieves. To jump the ignition on a recent car you will have to connect your jumper cable from the hot (ungrounded) terminal of the battery to the positive (battery) terminal on the coil. A jumper cable can be used to activate the solenoid or to bridge it and carry juice directly to the starter. Don't run far with the ignition jumper in place since it bypasses a resistor and the coil won't last long.

On many late model cars the ignition wiring is fancier and the above procedure won't work. A very handy gadget to carry in your car is a remote control starter switch with its own lead wires and built-in circuit breaker. Commercial models come with instructions covering use with various late model cars.

Troubleshooting the generator and voltage regulator is covered in Chapter 4.

LOCK OUTS

When you are outside the car with the door locked, and the key is inside, it can be pretty embarrassing. Try to borrow a wire coat hanger and bend the hanger part straight. Slip it through a vent window, if you left one open a bit, or through the rubber window molding, and try to catch the locking knob with the hanger hook. You may have to smash a vent pane to get your wire in.

EMERGENCY TIRE REPAIRS

Remember back in Chapter 5 we said that a bead breaker and bead expander were two special tools needed to get a tubeless tire off the rim and back on, so that it could be permanently patched on the inside. There's a way of getting around this requirement, however, and here it is.

The tools you will need are tire irons (or two pry tools such as a crowbar and a large screwdriver), a patch kit, a valve core wrench, a small bottle of detergent, and a pump to inflate the tire once it is repaired.

First jack up the car, remove the flat, and look for the puncture or listen for the hiss of escaping air. If the leak is not easily found, douse the tire liberally with a little detergent mixed with water, then look for the place where soap bubbles are forming. Remove any foreign objects from the tread with a pair of pliers.

Deflate the tire fully by loosening the valve core inside the stem with your little valve core wrench. Some valve caps have slots and can be used to unscrew the core in a pinch.

Use your regular car jack to break the bead. By putting the flat under your jack base and jacking the car so that the weight is on the tire bead, it will loosen at that point. Once the bead is loosened, you can stomp it down all around the tire with your old Army boots. Then pry one edge of the tire completely up over the rim, using two tire irons or similar pry tools. Pry the second edge up over the rim and the tire is off.

The latter step can be a real pain unless you have some way to hold down the tire firmly while you pry at it. It usually requires two people and a lot of grit.

A good patch kit will have a scraping tool, rubberized patches, and some type of cement. Scrape around the puncture area, roughening the surface so the cement can get a good grip. Then apply the patch, smooth it down, and let it dry. There are also hot patches which are applied to the surface and ignited with a match to form a secure bond with the tire. Either hot or cold patches will do the job.

Apply some of your soapy water to the tire bead to lubricate it and you are ready to remount the tire by prying it back on the middle of the rim. Constrict the tire by putting a chain or heavy rope tightly around it. This will ensure that it seats correctly on the rim when inflated.

As a safety precaution, *do not inflate the tire more than 35 pounds until the bead is fully seated*. Otherwise the tire could explode. Constrict the chain or rope enough to help the tire reseat itself correctly. Inflate to proper pressure as the final step.

A slow leak around the rim can be fixed by removing the valve core again, pouring some rim sealant into the tire, sloshing it around thoroughly, then putting the valve back and reinflating. Tire rim sealant is a standard accessory store item and you may want to carry it in your roadside kit.

Tubed tires are fixed in pretty much the same manner as tubeless ones. The bead on a tubed tire is easier to break and the patch goes on the tube. Or you can carry a spare tube or two in your emergency kit.

Since repairing tubeless tires can be an arduous and frustrating procedure, use a temporary plug inserted from the outside whenever you can.

The handiest air pump for field use is definitely not the "bicycle pump" type, which quickly produces a pain in the small of the back and lower regions. Instead, try the type that screws into a spark plug hole and runs on engine compression. An electric air compressor is still more convenient to use, but also bulkier and more expensive.

CHASSIS AND EXHAUST SYSTEM

Strange noises are often a tip-off that something has come loose under the car. It may be a dragging exhaust pipe, a shock hanging from one bracket, or a loose U-bolt on a leaf spring.

Lash up a dragging exhaust pipe with wire. Take a loose shock off if you don't have the nut and bushing to attach the loose end. Tighten leaf spring U-bolts.

If a spring is missing altogether or a tie-rod on the steering gear is bent use your ingenuity. Perhaps a sapling can be temporarily rigged up as a spring? High-strength safety wire and miscellaneous bolts, shackles, links, rods, and other connectors or fasteners are a vital part of your emergency kit if you are really traveling in the boonies.

Use caution when attempting makeshift repairs on the steering. The life you save may be paying your son's way through college.

ACCIDENTS AND FIRES

After a minor accident a fender pushed against a wheel may be the only thing immobilizing your car. Use a crow bar, tire iron, or piece of wood to pry it away. A hacksaw may be needed in some cases. You can also wire down a sprung hood, lash up a door, and replace a broken light bulb if the socket is intact.

Always turn the ignition system off after an accident to prevent fires caused by a loose wire short-circuiting. In the event a fire does start, grab the trusty fire extinguisher you always carry and put it out. You can also smother a fire with a blanket or piece of clothing. Never throw water on an electrical fire.

For smoldering wires the best remedy is to take off one battery cable as quickly as you can. Then replace or repair the wires.

There is always the danger of a gasoline fire on a car and it can flare up quickly. The gasoline in the tank can explode. The safest way to deal with a gasoline fire is to get far away from it as fast as your little legs will carry you and hope that fire damages are covered by your insurance policy.

TOWING

When all else fails, you may well find a good samaritan who is willing to tow you to a nearby garage. If he has a fixed tow bar, which clamps on to your bumper, on his car, towing is simple. A tow chain or rope is another story.

Towing with a chain or rope should be done for short distances only. Try to hitch the chain to chassis members or anchor shackles attached to the frame rather than bumpers which may prove inadequate. The chain or rope should be at least ten feet long so that the car being towed has room to brake. Clevis grab hooks at each end of a chain make the best connections.

When towing or being towed, it is advisable to proceed at slow speeds with the flashers on the towed car activated. The two drivers should have a simple system of hand or horn signals to indicate when either is putting on his brakes.

Manual transmission cars can be towed in Neutral with the engine off. Auto transmission equipped vehicles may have to be towed with their rear wheels off the ground.

Auto transmission cars from GM, Ford, and American Motors can be towed with the rear wheels on the ground for short distances only (10 miles or so) at speeds under 25 m.p.h. The transmission should be in Neutral. Chrysler Corporation cars can be towed a greater distance—about 100 miles. Cars towed for greater distances than recommended should have their rear wheels elevated or the driveshaft removed.

GM cars after 1968 should be towed with the ignition key on whether or not the rear wheels are off the ground. Many models in 1969 and all later ones have a system which automatically locks the steering and shift mechanism when the ignition is off. When the ignition key is missing, these cars must be towed with both front and rear wheels off the ground—on a flatbed trailer.

CONCLUSION

There is an old saying that all car troubles are caused by a loose nut behind the steering wheel. To keep from being that loose nut you should realize that proper maintenance is the key to avoiding roadside emergencies and is a vital part of auto safety. When you do break down far from home, use common sense about repairs which involve the brakes, steering, and tires.

9

Getting Unstuck

● Once you get stuck on snow, mud, or sand, don't keep spinning your wheels in a vain effort to get out. This is known as digging your own grave, since the wheels won't go anywhere but deeper.

The first thing to do is get out and look to see how badly you're stuck. Better yet, if you have a passenger, ask him to get out and see which wheel is losing traction. Of course, you are concerned with the drive wheels, which are the rear wheels only, unless you have a front-wheel-drive or four-wheel-drive car.

If you do not have a limited-slip differential (Positraction, Sure-Grip, or some other brand), then one wheel may be spinning madly while the other is totally immobile. Your objective is to get some traction for the spinning wheel.

With limited-slip you will go forward so long as either wheel has traction and only stop when both wheels are spinning. In this case you must improve traction for either or both of the wheels.

If your wheels are spinning on ice, it may be a simple matter of starting up gently. Start ahead in second gear (on a stick-shift) and use a gentle foot on the gas. Also try pulling the handbrake on a few notches. By limiting power you may be able to keep the wheels from spinning and to get going again.

On any surface you can probably get your car out by rocking it if it will move at all in forward or reverse. Rocking a car is accomplished by rapidly alternating between attempts to drive forward and attempts to reverse. With an automatic transmission, come to a complete stop before moving the lever between drive and reverse. If you have a standard transmission, put in the clutch and begin shifting while the car is still moving, but don't complete the shift until the second you stop.

When you rock a car it will usually begin to free itself and take you a little farther each time until you're unstuck. But don't rock the car too long, especially with an automatic transmission. Rocking puts a big strain on transmission, driveshaft, and tires.

Along with rocking, you can try using traction improvers. One of the best is sand, which can be sprinkled in front and back of the drive wheels. The sticky stuff sold in aerosol cans as a traction improver also works well when you spray it directly on the drive wheels. If you don't have anything to put around the drive wheels for traction, look on the shoulder of the road for gravel or a little earth you can dig.

Instead of sand, some people carry a sack of rock salt or calcium chloride in their cars during the winter. Salt is effective, but it will rust out the inside of your trunk compartment mighty quick if the sack bursts.

Once you are too far in for rocking and sand to be effective, it's time to get out the old shovel. Shoveling works well on snow, but sand or mud may not offer a solid surface even if you dig down deep.

Those little Army surplus folding shovels are a pain in the neck. The job will go a lot easier and quicker with a short, D-handled shovel which has a substantial scoop at the end. Shovel in front of the drive wheels and in back of them if you want to try rocking. You might also have to shovel in front of the car if the snow is deep enough to cause a sizeable obstruction.

On sand or mud it is usually best to try jacking the rear up and putting boards under the wheels. The best type of jack is the Hi-Lift since it can lift the car as much as four feet. This is enough to pull the wheels clear of gooey mud and is a great help if the vehicle is "high-centered" (resting on the oil pan, axle, or some other under-component rather than the wheels). Other jacks can do the job also, with bumper types being the most versatile.

Along with boards under the wheels, you can use stones, burlap sacks, chicken-wire, or absolutely anything you can scrounge that looks like it might do the job. Don't neglect the spare tire and wheel, which can sometimes work nicely. Rubber floor mats are good also.

Because the base of most jacks is narrow, they need support to keep from sinking into soft material. Make a jack base from a piece of exterior plywood about a foot by a foot-and-a-half and at least an inch thick and keep it handy for emergencies. If the jack base has holes through it (as the base

Hi-Lift jack is most versatile for de-ditching and emergency winching chores. (Hi-Lift Jack Co.)

on a Hi-Lift does), try bolting the plywood to it for greater stability. Use wing nuts so that you can attach and detach the plywood quickly.

Once you have boards or sacks under the wheels, you will probably have all the passengers out pushing while you try to get going. Just make sure they are cautious about where they push from, since a spinning wheel can throw a board back at someone's leg with a lot of force.

Another technique to get out of a hole is to jack up the wheels and then push the car sideways so that it topples off the jack. The higher the car is,

Car can be pushed over on jack to move wheels out of ditch.

the easier it will be to push over and the farther it will go. The Hi-Lift jack is most effective and is rugged enough so that it won't be damaged by this operation. This is one of the most effective ways of getting out of a rut or pothole.

If you jack a vehicle which is on an incline, it may start rolling as soon as it's free. To prevent this from happening when you don't want it to, put chocks under the wheels.

NOW STOP. If you haven't gotten out by this time, take a break to think it over. Wipe the sweat off your brow and start thinking about whether you can get some friend to come around and help pull you out with his car. If you can, remember to hook your tow chain around the frame members of both cars, since bumpers can be weak.

A lot of cars don't have any convenient place to attach a chain. You could always buy a trailer hitch or pintle hook for this purpose, however a cheaper method is to look for a place you can permanently attach an anchor shackle to the frame. Then your tow chain hook can attach to the anchor shackle. (Don't know what an anchor shackle is? Just ask the clerk at your friendly hardware store.)

Whether you are being pulled out or trying to get out under your own power, be careful to keep your wheels straight. Wheels crimped to one side or another have to push too much snow or mud in front of them. Even if another car must pull you at an angle, keep your own wheels straight.

Should you be unable to get another vehicle to pull you out, your only alternative is winching. Fortunately you can do this with a Hi-Lift jack, provided you have extra cable or chain.

The clevis at the top of the jack can be fastened to a nearby tree or other solid object with a long chain. The lifting part of the jack can then be fastened to your car with another chain. Now the jack is horizontal and is applying its pulling force sideways. It can move three to four feet at a time before the lifting part of the jack has reached its apex and you have to get out and re-rig it with a shortened chain. Each time you re-rig, you will probably also have to chock the wheels to keep the vehicle from slipping back into its hole. This method is effective but slow.

A piece of three-eighth-inch aircraft cable about 100 feet long is cheaper and much lighter than a chain of similar length. The only drawback to cable is that it can fray easily, so use chain to make end connections around a tree or other fixed object. Have the cable rigged with slip hooks at each end and attach it to the chain at one end and the jack at the other. For attachments between jack and car, use your tow chain.

Of course, a power winch or hand winch of the "come-along" type is easier to use than a Hi-Lift jack for these purposes. You can get about 50 feet of cable on a hand winch and 200 feet or more on a power winch. All this length is straight pulling without having to stop every four feet and re-rig. The only drawback to a winch is the expense.

Be sure to always stand in a place where you are protected from the wire cable when winching. A breaking cable can snap back with considerable force.

If you attach your rig to a tree and, instead of moving the car, it pulls the tree down on top of the car, you miscalculated. In such a case you will need not only a tow truck, but also a good body man and maybe even an ambulance. So be conservative.

What if there's no tree or other fixed object to which you can attach your rig? In that case you will have to make your own fixed winching point through use of a "deadman." The simplest and cheapest deadman is made from a two by four with the biggest eyebolt you can find bolted through the middle of it. If you drive a lot on bad roads and have a winch, keep the two by four in your trunk compartment at all times.

When you need to set up a deadman, begin by digging a slit trench the length of the board in front of the car and parallel to your front bumper. Make it as far out ahead of the car as your rig will stretch. Dig another slit trench perpendicular to the first one and facing toward your car. This

Winching with a "come-along" and deadman. Top: the "come-along."

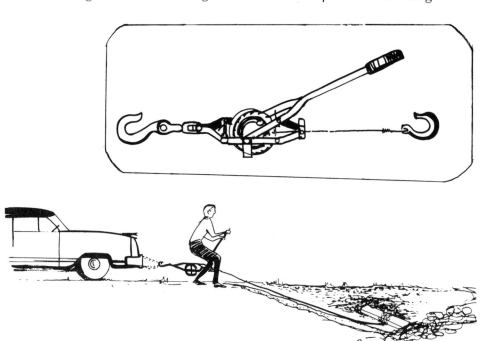

should meet the first trench at its center. The second trench should be approximately seven feet long with a gradual upslope so that it reaches the surface in the direction of your car. A pickaxe will make digging these trenches in hard ground a lot simpler.

Place the board in the trench parallel to your bumper and attach the chain to the eyebolt. Then run the chain out through the perpendicular trench. When the chain is attached, bury the board and stomp the dirt down on top of it. The better the board is buried and the deeper it is, the less chance it will come out when you start pulling with your rig.

Now attach the chain to your rig and start pulling. It's a lot of work, but worth it when you're stuck in the middle of nowhere.

If you do decide to spring for a winch, look over the available models carefully. The various types of power winches are handy for four-wheel-drive vehicles, especially those driven in desolate areas as part of club activities. It's usually simple for a man with a winch to pull out another club member who gets stuck.

For pulling yourself out of trouble with a vehicle which is not designed for off-road travel, the come-along type of winch has some advantages. Since it is not fixed to the vehicle, it can pull from either the front or rear with no problems. Also, being manual, it will work just as well when the engine isn't running and the battery is dead.

Depending upon the pulling power and cable length, a good come-along, such as the Lug-All, will cost from $40 to $150. The two-ton pulling power, which is maximum for most hand winches, will almost always do the job.

The Lug-All can also be rigged as a hoist and used to lift just about anything you can think of. It's also handy if you want to play tug of war with an elephant.

Well, that's about it. If you are reading this chapter while stuck out in the mud in the middle of nowhere and nothing has worked yet, you're in trouble. Better start looking for a farmer with a good heavy tractor.

10

Major Repairs

● My interest in cars began at a tender age when I was given a book about the adventures of a little red automobile named Chuffer. Chuffer had a series of owners who made him haul heavy loads and forgot to change his oil. They neglected him so much that his engine began to wheeze and his chassis started to clunk and he eventually wound up on a junk pile. At least that's the way I remember it.

Then, as you might have known, Chuffer was rescued by a heroic type who recognized his sterling worth beneath the coat of rust. With a rebuilt engine and some other unspecified repairs, Chuffer lived happily ever after.

As I grew older, I sometimes wondered whether Chuffer really did live happily ever after like the book said. Or did he drop his transmission after a couple of hundred miles more and wind up on the junk heap again?

Despite these occasional doubts, the Chuffer story has probably been the root of my lifelong and incurable optimism about cars. As I continued to grow up, I graduated to reading sports car and hot-rod magazines which offered a different brand of fairy tale. They told about guys who slipped rip-roaring, hemi-head engines into Chuffer-like jalopies, and presumably they also lived happily ever after, drag racing and opening up their muffler cutouts as they cruised past the drive-in.

None of those articles said anything about winding up with a cellar full of esoteric wrenches and three rusty parts cars moldering away in your backyard. They forgot to mention the fact that Chuffer invariably needs a new transmission and a stronger rear end and an oversized cooling system and better brakes to go along with his powerful new engine, and that cars cost a whole lot more when you buy them piece by piece.

Above all, those magazines left me with the impression that automobiles were sort of like overgrown erector sets in which you only had to slip tab A into notch B and connect bolt C through hole D for everything to fit together perfectly. They passed lightly over the fact that erector sets are a lot tougher to take apart and assemble when you are lying on your back on a cold concrete floor and groping overhead to reach a bolt which has been cunningly tucked away behind obstruction E. They didn't say that it complicates matters no end when the parts of your erector set get worn, bashed in, or tenaciously rusted together. Furthermore, they never intimated that you could ham-fistedly drop screw F down niche G where it would be lost forever.

The lesson I hope I have learned is not to exceed my modest abilities as a mechanic. Doing minor repairs and maintenance makes good sense from an economic viewpoint. This is not necessarily true for major repair work requiring specialized tools and long experience in using them.

Often a knowledge of where to go for parts and services, plus an acquaintance with fundamental repair procedures, will result in substantial savings on the cost of professional repairs.

At other times the best alternative may be to do part of the repair yourself and have the critical portion done by a mechanic who has the skills and tools that you lack.

There will come a time, however, when the weather is warm and your enthusiasm runs high. You will have a place to work, a friend or two to help out, and an ample supply of beer in the fridge. Most likely you will be fixing up a second or third car, so there's no deadline for getting it back on the road. These are the ideal conditions for attempting your first big job, and you may find the experience so enjoyable that visions of rescuing some rusty old Chuffer from the junk heap will afflict you for the rest of your life. Don't say I didn't warn you.

Now let's begin with a rundown on professional repairs and trade practices. Then, after covering a bit on shopping for parts and some tools you'll need to purchase or borrow, we will go over the major repairs you can carry out most successfully on your own.

WHERE FOR REPAIRS?

When you get sick, the first person you are apt to go to is your family doctor. He may refer you to a specialist, or send you to a hospital if the ailment is serious, but he still serves a very useful function. You can rely on him for advice, diagnosis, and referral.

A family mechanic can fill a similar need when your car is ailing. In the long run it will pay to buy your gas and tires at a small garage run by an experienced mechanic rather than at a run-of-the-mill gas station or discount center.

The small independent mechanic can handle any minor repairs which you don't choose to do yourself, and some larger jobs too. For instance, he should be perfectly capable of repair work involving the replacement of complete units such as an engine or transmission. The independent owner/mechanic will also go to the trouble of locating used or rebuilt parts while larger service centers are not often so obliging.

Just as you cannot expect a family doctor to do brain surgery, the small mechanic, who is a generalist rather than a specialist, also has his limitations. The next logical place to seek repairs is the service facility at a franchised dealer.

The new-car dealer carries a large stock of parts and has enough mechanics so that each can specialize in a particular aspect of auto repair and get a lot of practice doing so. The dealer also has the direct support of the factory, which supplies him with service bulletins, catalogued tools, and training for mechanics.

The dealer is usually best equipped to service the more recent and complex assemblies on a car, such as fuel injection, emission-control systems, climate-controlled air-conditioning, anti-skid brakes, and electronic ignition systems. He should have an advantage over the independent garage when it comes to doing really rough jobs like complete rewiring on a car damaged by fire. Also dealers often have parts on hand which are not available at local supply stores, such as heater cores, power steering pumps, disc brake assemblies, and factory rebuilt engines or transmissions.

The large independent garages do have some clear advantages over franchised dealers. Usually their overhead is lower and they can buy parts from local suppliers at prices which reflect a highly competitive situation. Thus, independent garages can frequently undercut dealer's prices on the more common repairs.

Since the large independent garages have no customers who automatically come to them for service after buying a car, they have to hustle more for their business. While the quality of their service is probably about equal to that of franchised dealers, on the average, there are fewer really bad ones.

Then there are the specialty shops. Some of them offer services which are not available at most garages, such as radio repair, welding, or upholstery. Body and fender shops, auto painters, radiator repair facilities, engine machinists, and front-end shops also offer worthwhile specialties which merit consideration.

The franchised specialty shops advertising automatic transmission repairs at budget prices, or the low-priced engine rebuilders, brake specialists, and pseudo-diagnostic centers are another story. Under one guise or another they really sell inferior parts or replacements which may be unnecessary. The diagnostic services and "free" checks are a gimmick to lure unwary customers. And their employees are trained to replace entire units rather than to do skilled diagnostic work.

There are legitimate diagnostic centers which offer valuable services and specialty shops which advertise genuine bargains. Separating the sheep from the goats is a task which requires no particular knowledge of auto mechanics, just a little common sense.

A final type of repair facility is the service center operated in conjunction with a department store or auto accessory store. These centers usually do minor repairs and install items purchased at the store. They are rarely equipped to do anything more.

It is surprising that some of the most barefaced rackets in the auto repair business are practiced at service centers that are allied with quite respectable stores. Apparently the service managers and employees are paid according to the amount of business they do, and there is little effective supervision.

Which brings us to the next topic, honesty or the lack of it.

GROUND RULES

If there is a widespread dissatisfaction with the quality of auto repairs, the main reason is the high cost of labor in America today. For those who can afford it, there are speed shops where engines are "blueprinted" to specifications closer than the factory ever used, and custom painters who lovingly apply endless coats of hand-rubbed lacquer. The rest of us must be content with some lower standard.

Auto dealers and most independent garages base their labor charges on manufacturers' flat rates. The hourly rate for labor is set by the garage, while the amount of time each job should take is based upon the manufacturer's flat-rate manual, revised annually.

The flat-rate system is designed to protect the consumer. Without it, an inexperienced mechanic who took ten hours to complete a repair could charge more than a skilled one who finished in three hours.

The flat-rate system does have one outstanding flaw. Many mechanics are paid a percentage of the flat-rate fee for each job rather than a straight salary. The faster they work, and the more shortcuts they take, the higher their profits will be.

With this "piece-work" system based on flat rates, a mechanic makes less on work done under warranty, so warranty repairs are often done perfunctorily or left to unskilled trainees. Certain other types of repairs also suffer. For instance, tune-ups are done swiftly with no time taken for the minor services—a dab of lube here or a bolt checked for tightness there—which make all the difference.

There is probably no more flagrant dishonesty in the auto repair business than in any comparable trade. It is simply that there are more opportunities to cheat. The industry is highly competitive, its growth has outdistanced the channels for training and apprenticeship, and the standards which have been established for repair procedures still contain many gray areas.

Some of the blame for the mediocre level of auto repairs today undoubtedly belongs to the manufacturers, their dealers, and the trade associations. However the heart of the problem is really the consumer. If a majority of people purchased cars on the basis of their serviceability and frequency-of-repair record, rather than because of styling and power, manufacturers' policies could change overnight. And if more people selected a dealer because of his reputation for service, rather than on the basis of advertising or a slightly bigger discount, then the quality of repairs would surely rise.

Of course, you and I are the exceptional cases. At least it's nice to think that we are. As the elite among consumers, we are willing to pay for quality work at an honest price. The question is, how can we be sure that we are getting what we pay for?

Parts and labor charges are easy to check. You can compare the price of the dealer's factory parts with those available at a professional parts supply house by a quick phone call to each. Then the mail order price for equivalent parts can be checked in catalogs like the comprehensive one put out by J. C. Whitney.

You will almost always find the mail order or department store price is less than that of a dealer or parts house. The difference is mostly overhead, but is sometimes reflected in the quality of the merchandise. Since most garages will not make repairs unless they supply the parts, the difference in price is a premium you pay for their services.

Flat rates for labor can be checked in a flat-rate manual which lists the amount of time (in hours and fractions of hours) that each repair should take on a specific car. If you do not have access to a dealer's book, you can buy a current compilation of flat rates (American cars plus VW and Toyota) from Irving-Cloud Publishing Co., Lincolnwood, Chicago, Ill. 60646 (price: $2). Flat rates are also given in some repair manuals.

While parts and labor charges can be readily compared, it's a lot tougher to determine whether a job is diagnosed correctly and carried out in an honest and competent manner. There's no way you can be 100 per cent sure all the time, of course, but you can take a giant step in the right direction by following a few simple ground rules.

When you leave your car at a garage, always write out your instructions to the mechanic and keep a copy. If there is any doubt about what's wrong with the car (and there almost always will be), just state the symptoms as clearly and completely as you can. Written instructions are not only legal protection, but also will help prevent garbled communications between the service manager and the mechanic who does the work.

Ask for a written estimate before authorizing any major repair. Be wary of garages which refuse to make any estimate before dismantling a faulty component. While it is true that the full extent of a necessary repair may not be known before disassembly, it is always possible to give a rough estimate with an explanation of problems which may necessitate additional costs.

Read your shop manual to get an idea of the steps recommended in carrying out the job. The object is not to be a know-it-all who tells the mechanic what to do, but to grasp what is being done and why.

Try to be present while the work is in progress, or at least at the critical point. Some garages will not let customers in the repair area, supposedly for insurance purposes. The real reason is that supervising customers often make pests of themselves, and too many on the floor at one time can lead to chaos. This rule should be relaxed if you are polite and keep out of everyone's way. At the very least, you should be allowed entrance while the faulty parts are removed and the new ones installed.

You are also entitled to an explanation of why a particular component failed. Sometimes the answer is normal wear or hard driving. When an abnormal failure occurs, however, an effort should be made to ensure that the root cause is determined. For instance, if the clutch is giving trouble and oil is found on the clutch plate, no mechanic who is worth his salt will install a new clutch plate before trying to determine where the oil came from. Perhaps a rear main bearing seal is leaking and it must be repaired at the same time the clutch plate is replaced or you'll be back again with the same problem.

Garages have a powerful legal weapon in the mechanic's lien which allows them to keep any car until the bill has been satisfied. Even if the charges are dishonest, this must be proven in court before the car is released. Your best defense is to choose a garage that has a reputation to maintain, give written and explicit directions prior to any repair, and always insist on a written estimate with an upper limit.

TOWING A DISABLED CAR

All your acquired wisdom about choosing a garage can be futile when you break down on the road and have to be towed in. The service stations on thruways tend to be rapacious, so it's a good idea to consider getting a tow home, or at least to a garage of your choice.

Towing with a chain is okay for short distances (see Chapter 8) but definitely not recommended for longer trips or on freeways and turnpikes.

A handy device for these emergencies is a universal tow bar which hooks tightly to the bumpers of both the tow car and the vehicle being towed. It is not necessary that the tow car have a trailer hitch ball, although there naturally are hitches made to be used with one. Either type of hitch should always be used with a safety chain.

On most roads it is legal to tow a car with the proper hitch during daylight hours provided that a sign saying "Car in Tow" is affixed to the back of the rear vehicle. In some cases it is necessary to have the type of hitch which includes cables to activate the brakes and stoplights of the vehicle in tow. At night the emergency flashers on the rear car should be used when the two vehicles are traveling very slowly.

An adequate towing rig can be had for less than $40, or about $90 for a heavy-duty bar with brake cables.

When you have no tow bar and no one to tow you, don't despair. Both a tow bar and towing vehicle can be rented. Although I understand it is current U-haul policy not to rent a tow bar unless a truck is also rented, universal tow bars are still available from some other rental agencies.

Most owner's manuals give long-distance towing recommendations. Frequently the car's driveshaft will have to be disconnected when very long distances are contemplated. Practically any garage can perform this service for a small fee.

For distances exceeding 300 miles, transport on a railroad flat car is a possibility. It costs less than the price of trucking a car and is often easier to arrange. However you will have to tow the car to and from the station.

Getting a tow home is an ignominious way to end a vacation, but it can be a lot more pleasant than waiting around in a strange city and then getting hit with a whopping repair bill from some dubious gas station.

PARTS

Whenever you can provide your own parts, either because a garage is permissive or when you're doing the work yourself, there's money to be saved in comparison shopping.

In Chapter 3 we discussed the parts situation and determined that rebuilt units were usually a good choice. They should last about as long as the brand-new stuff and save you enough folding green for a night or two on the town.

On some rebuilts, such as fuel pumps, voltage regulators, and even clutch pressure plates, the price differential is usually not all that much. So it may pay to purchase them new if it makes you feel more secure. Greater savings are possible on rebuilt starters, generators and alternators, carburetors, water pumps, power steering pumps, engines, automatic transmissions and manual transmissions.

You can also buy rebuilt short blocks, which are engines without the cylinder head and auxiliary parts, and torque converter components for automatic transmissions.

Used parts are more risky. Body parts can safely be purchased from a wrecker because defects are readily spotted, but high-wear components such as clutches, brake cylinders, carburetors, and steering boxes are not really worth the gamble.

The list of used parts that are worthwhile looking for includes the starter and all small electric motors (windshield, power seats, etc.), radiator, heater, air-conditioning assemblies, seats, instruments, driveshafts, differentials and rear axles, springs, wheels, and anything else that's cheap enough and doesn't have a lot of internal moving parts to wear.

Don't automatically assume that wrecking yard prices are the best bargain. Wreckers have a lot of overhead to carry and some of them charge whatever they think the traffic will bear. In extreme cases they have been known to charge more for a part than it would cost if purchased brand-new from the dealer.

Sometimes it is possible to buy a whole car from a private party for the same price a wrecker would charge for the engine or transmission alone. And if the car is still running, you can get a better idea of what shape the units are in. On the other hand, you won't get any guarantee with the old junker and you'll have to find a place to store it which won't rile the local zoning board.

When you are in no particular rush for parts, mail-order prices are often lowest. However, the cost of shipping heavy items can be considerable. One way in which you can often save a few dollars is by calling up the express company and asking them to hold the shipment until you pick it up. This will prevent any surcharge for home delivery.

It is not always easy to tell which parts fit your car. Catalogs are often ambiguous and sometimes inaccurate. When there is any doubt, it's best to inquire before ordering. Send a self-addressed stamped envelope and leave room on your letter for an answer to be scribbled because parts suppliers are not in the habit of writing formal replies. Always list the chassis and engine number of your car since manufacturers do make modifications during the model year.

It is best to send a money order or certified check in order to speed delivery. If you send a regular check, many suppliers won't ship until it clears the bank.

SHOP MANUALS AND TOOLS

When you are doing major repairs, a shop manual giving step-by-step procedures for your car is pretty essential. In most cases you have a choice between the official factory manual and privately published service manuals. (A list of manuals may be found in Appendix E of this book.)

Manufacturers' service manuals are often more detailed, but they are aimed at the professional mechanic and can sometimes be confusing. For instance, factory manuals frequently refer to a special tool by its catalogued part number without indicating that some combination of ordinary hand tools can be an effective substitute.

Quite a few service manuals for foreign cars are published in England. England is the place where people work on their saloons and go off to the pub rather than fixing their cars and heading for the saloon. They strangle their carburetters instead of choking their carburetors and always remove set screws (bolts) with a spanner. Batteries are often positively earthed (grounded) and engines have gudgeons, which I always thought were a form of blind fish inhabiting the water jackets around the cylinders, but which turn out to be nothing more mysterious than piston pins. With a little imagination it's not too difficult to translate English into American.

After buying a service manual, your next investment should probably be a torque wrench. A torque wrench is a type of socket handle which can be used with all sockets and adapters. It measures the twisting force applied to a bolt and prevents overtightening.

Shop manuals give torque specifications for just about every bolt on a car —even the wheel lug nuts. Now you and I know that no one ever uses a torque wrench to tighten wheel lugs, but when it comes to head bolts, main bearing cap bolts, and many other critical parts of the engine, transmission, and power train, accurate torquing is the only way to go.

A torque wrench reading from 0-150 ft. lbs. (not inch-pounds) with a half-inch drive is most useful. Inexpensive bending-beam-type torque wrenches ($10-15) are accurate enough for most purposes, but the more expensive preset wrenches are a lot handier to use in tight places.

Bolts actually stretch a little when they are tightened. When a bolt is overtightened, it elongates too much and grows weaker. At some point it will break. This is why bolts are manufactured in various strengths and torque ranges are specified. Always replace a bolt with one of the same grade or better.

Sooner or later you'll manage to break a bolt and be faced with the problem of extracting the stub. The usual method is to drill a hole in the middle (with an electric drill which should be high up on your list of essential tools) and screw in an Ezy-Out screw extractor. A set of Ezy-Outs and a tap wrench to turn them should be within your budget.

When the inside thread of a bolt hole is damaged, you will need to tap it out and install a thread insert such as a Heli-Coil. Unless you want to invest in a tap and die set plus other special tools, this is a job best left to an automotive machine shop.

Another pretty essential tool for the home mechanic is a good, heavy vise. You can pound on the anvil part, use the jaws for bending sheet metal, and even apply pressure to parts which are press fit.

There'll come a time when you have to take an engine or transmission out and you can't round up six strong friends to assist with the job. That's when you need a winch, block and tackle, or chain hoist on engines which are removed by lifting. In addition, an engine lift cable is needed to attach to the block and it should have a control ring to adjust the lift angle. Engines which are removed by lowering (VW and Corvair, for instance) can generally be slid out with the aid of a hydraulic floor jack. An alternative is to fashion a substitute out of a couple of inexpensive scissors jacks bolted to an old dolly.

Welding is a skill which any amateur mechanic can use once in a while. Although the more challenging jobs require quite a bit of experience, almost anyone can learn to do some simple brazing or cutting in a few lessons.

Preset torque wrench. (Owatonna)

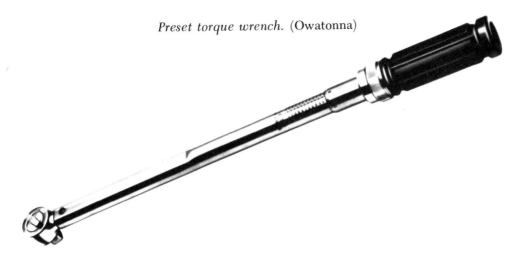

*Broken threads inside bolt hole or spark plug hole
can be renewed by insertion of Heli-Coil.* (Heli-Coil)

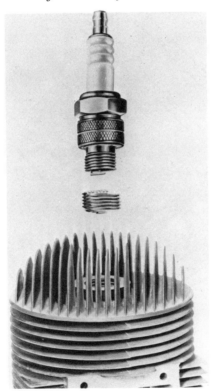

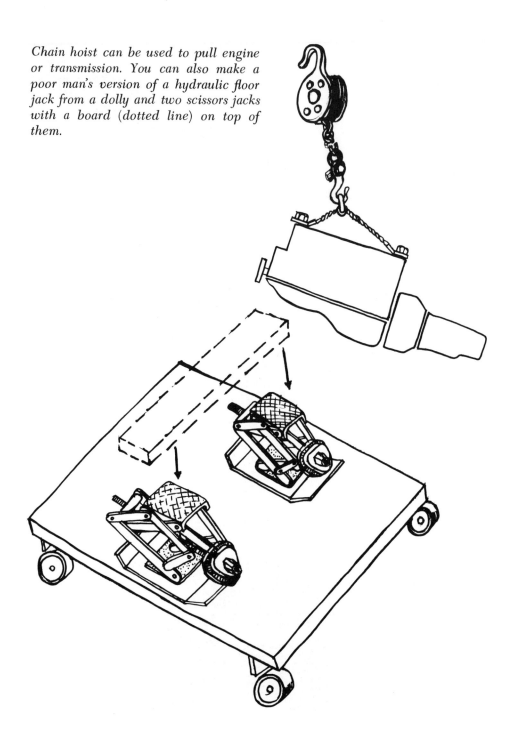

Chain hoist can be used to pull engine or transmission. You can also make a poor man's version of a hydraulic floor jack from a dolly and two scissors jacks with a board (dotted line) on top of them.

The strongest welds are made with an electric arc welder, but an oxy-acetylene torch is much better for cutting and general-purpose brazing. (Brazing is the process of joining two metals with bronze or alloy filler rod which melts at a lower temperature than steel. In welding, two similar metals are directly heated until they flow together.)

An oxy-acetylene torch set with tank valves and various tips will cost about $125. The oxygen and acetylene tanks are not included since they are owned by the supply company, like bottled-gas tanks. There is a demurrage fee of a few dollars a month as long as the tanks are in use.

Since welding equipment is pretty expensive for those who will make use of it only once in a while, the low-cost solid oxygen outfits are becoming increasingly popular. They burn solid pellets of an oxygen-generating compound and use propane fuel rather than acetylene. A solid oxygen rig won't produce the heat of an oxy-acetylene torch (about 6000° F.), but it will get hot enough to do small brazing jobs and cut through thin metal plate. This is adequate for many smaller jobs.

A versatile tool for removing timing gears, various types of bearings, and many other parts which are press-fitted in place is a set of pullers. There are both screw-type and slide hammer (impact-type) pullers, and they may have two or three jaws and hook in place over external or internal lips. Companies like Snap-On sell modular kits with components which fit together to make up pullers of various types.

Finally there are a host of more specialized tools which are associated with particular repair jobs. You won't want to invest a fortune in equipment you may use only once in every few years, but some of the less-expensive items may prove to be a sound investment.

Milling machines and precision micrometers needed for doing a complete engine overhaul are beyond the average budget. However the simpler hand tools, such as piston ring compressors and valve spring clamps, are more feasible and will enable you to do a portion of the work and leave the rest to an engine machine shop.

Similarly, you can buy some body hammers, dollies, and picks in order to straighten a fender or repair minor dents on your car. When a bad bash occurs, however, you had best leave it to a body and fender shop which has heavy hydraulic pullers and other equipment necessary for more advanced work.

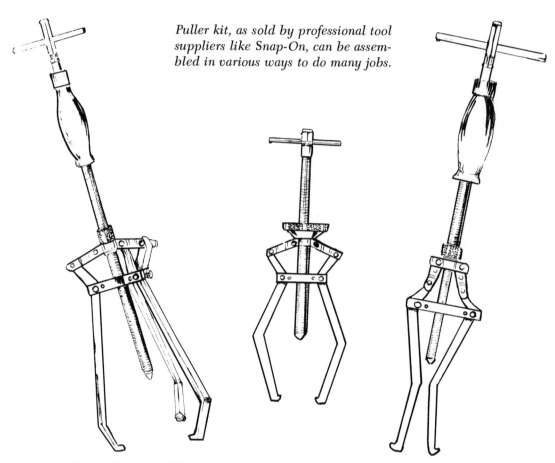

Puller kit, as sold by professional tool suppliers like Snap-On, can be assembled in various ways to do many jobs.

The technique of doing part of a repair yourself and getting professional help for the critical portion can work well if you know a cooperative mechanic.

TECHNIQUES

There are some working procedures applicable to almost any major repair and we'll go over them here.

The aspect of repair which is frequently most troublesome is the initial diagnosis. How do you know which part (or part of a part) is faulty?

One way is to replace it in toto and see if the car runs better. Obviously this is not the method of choice.

Service manuals do give test procedures for electrical equipment, engines, automatic transmissions, and some other components. There are also books such as *Glenn's Auto Troubleshooting Guide* and *Motor's Automobile Troubleshooter* which offer a detailed guide to diagnostic procedures. The trouble with troubleshooting guides is that they can't make you an instant expert any more than a French language dictionary can enable you to become fluent in the language overnight. They may be a valuable aid, but not a substitute for experience.

Diagnostic centers with electronic test equipment are a partial answer at best. An experienced operator may interpret the test results meaningfully,

Most useful body tools include (counterclockwise from top): tinsnips (A), body file (B), special hammers (C, D), pull rod (E), and various dollies (F, G, H), for shaping metal. Spray gun (I) can be rented.

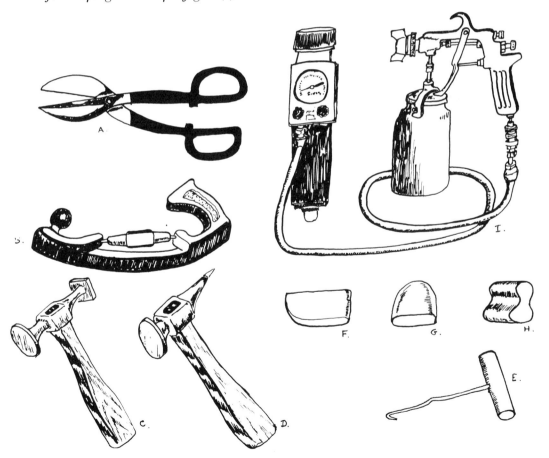

yet there are certain subjective factors which diagnostic tests never consider. For instance, a good engine man might advise you to replace your motor rather than repair it after considering the car's age, the availability of rebuilt parts locally, and the cost of labor for a complete overhaul. There are no test machines that can match his performance.

The ultimate answer is to consult an experienced mechanic whenever you are faced with a major job and aren't sure of what has to be done. Don't ask a gas station attendant for free advice or go to your Cousin Henry for his opinion because he once fixed a washing machine. Go to the best man you can find and expect to pay for his advice. You'll wind up saving money and learning a lot besides.

Once you are set on what needs to be repaired, don't rush out to buy the new parts before you disassemble the car and have a good look at the defective ones. Otherwise you may find yourself stuck with parts that aren't needed. Do check and be sure that the parts you are likely to need can be had, however, or you could end up completing the job next Christmas.

Choose a place to work which is away from neighborhood kibitzers and as comfortable as possible. The better your working conditions, the more you will be apt to take your time and do a thorough job.

Proceed to work methodically. You will find it helpful to think out a job beforehand and even to make a list of steps to follow.

Examine each part carefully before you attempt to remove it. Make a little sketch of how it fits on if you think you will have trouble remembering. Use little tags or pieces of masking tape to label disconnected wires, hoses, or mechanical linkages.

After you remove a part, put all bolts, washers, and screws back in the right holes so you won't forget where they fit. When you cannot put bolts and fasteners back on immediately, keep them in separate envelopes or boxes which are clearly labeled. An old egg carton with dividers is handy for small parts.

As an indication of how certain parts fit together, scribe a mark where they line up. Or use a punch to make two little indentations which can be aligned.

Avoid forcing parts that won't come off readily. First look closely to be sure you have unfastened all bolts. Then see if you can spot snap rings, bushings, pins, tabs, Woodruff keys, or other fasteners you are not familiar

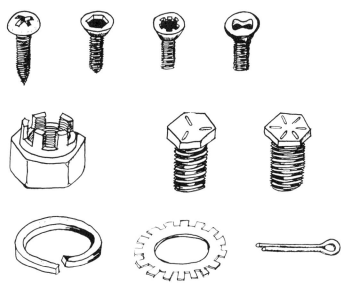

Fasteners you should recognize are (top row, left to right) Phillips, Allen, fluted and clutch head screws; (center row) lock nuts and bolts of more than common strength with grade markings on head; (bottom row) lock washers and cotter pin.

with. Parts which are stuck in place with gasket sealer can be loosened by tapping with a rubber hammer or piece of wood once the bolts are re-moved. Assemblies which are press fit into position may require pull-ers. Those that are corroded in place can be freed with penetrating oil, the application of heat from a torch, or even by using a punch and hammer to drive a nut around or chisel through it. When your ingenuity is ex-hausted, ask for advice before you reach for the sledgehammer and crowbar.

Often you will decide to clean greasy parts before reassembly. If you are using a caustic solvent, be sure you don't put rubber or plastic parts into it by mistake.

When disassembly is complete, drive (or hitchhike if you only have one car and you're working on it) down to the parts store and bring the defec-tive parts along so you can match them up with replacements.

Reassemble carefully, following the disassembly sequence in reverse. Spread a cloth under your working area so that small parts you drop won't roll away to some dark corner. Start bolts in their holes by hand to avoid stripping threads. When bolt holes are inaccessible, a magnetic screw starter is a handy aid. Hang up an extension light close-by to illuminate your working area.

Try to test whatever components you can before they are fully reassembled. That way you won't have to loosen everything again if they fail to work.

Road test the car as a final step and recheck all bolts for tightness.

ENGINE

Now let's turn to some specific procedures.

A complete engine overhaul will probably cost more than installing a rebuilt engine and will be more chancy. Therefore you should normally do no more than a valve or ring job at most, or perhaps a small amount of work on the bottom end.

You are taking a calculated risk by doing half a job. Vital parts of the engine, such as the crankshaft and cam, will not be touched and they cannot be expected to last forever. However you are renewing the parts which fail soonest at a relatively low cost. It is likely that you will be able to keep the car running long enough to more than repay your investment.

If you have decided that only a valve job is necessary, the job should cost $60 or less for machine work on a modern overhead valve engine and can be completed in a couple of days.

You are going to remove the head from the engine and bring it to an automotive machine shop, or engine "jobber," who will grind and reseat the valves. If necessary, he will mill the head, install oil seals, and replace weakened valve springs. Then you will reinstall the head, using new gaskets.

Your service manual should give you the exact steps necessary to remove the cylinder head (or heads on a V-8). Generally you will disconnect the battery cables, drain the cooling system, disconnect the throttle linkage and fuel line, and remove the valve cover or covers. The intake manifold and carburetor will be removed as a unit. The exhaust manifold can either be unbolted from the exhaust pipe and removed with the head, or it can be detached from the head, whichever is easiest. On some cars the alternator must also come off to give clearance for head removal.

The next step is to loosen the head bolts. If they are of unequal length,

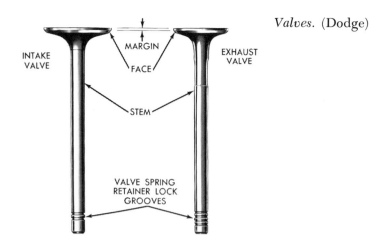

Valves. (Dodge)

INTAKE VALVE

MARGIN

FACE

EXHAUST VALVE

STEM

VALVE SPRING RETAINER LOCK GROOVES

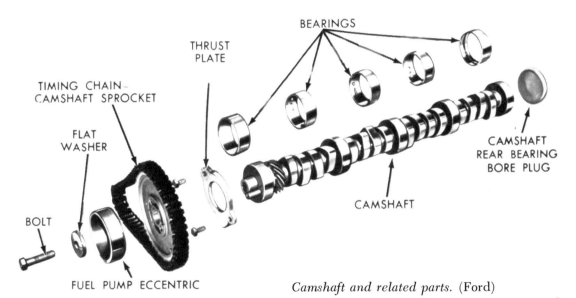

BEARINGS

THRUST PLATE

TIMING CHAIN – CAMSHAFT SPROCKET

FLAT WASHER

BOLT

FUEL PUMP ECCENTRIC

CAMSHAFT

CAMSHAFT REAR BEARING BORE PLUG

Camshaft and related parts. (Ford)

be sure you mark which goes where. Then remove the rocker assembly and pushrods.

On many engines it is possible to remove the rocker assembly without changing the valve clearance. However, if the car has hydraulic lifters and adjustable rockers which must be removed individually, note the number of turns it takes to unscrew each adjusting nut, and keep the nuts and rockers in order. That way valve adjustment will be easy when you re-install the head.

When the head is stuck to the block and won't lift easily, don't try to pry it off. Tap the head with a soft-faced hammer; or lever at some point on the outside of the head rather than between the head and block.

After the jobber is done with the head, you can clean carbon and lead from the combustion chambers with a wire brush chucked in an electric drill. Also scrape gasket material from the mating surfaces of the block and head. Turn the engine manually to bring each piston top to its apex and scrape it free of deposits.

When re-installing the cylinder head, you must use a new head gasket and follow manufacturer's specifications on the pattern and torque limits when re-tightening the head bolts.

After reinstalling the head, adjust the valves. Procedures for valve adjustment on cars with mechanical lifters are given in Chapter 6. Hydraulic lifters of the adjustable type are set for zero clearance between the valve stem and rocker, or rocker and pushrod. The adjustment screw must be tightened slowly, a quarter turn at a time. When clearance is zero, the adjusting screw must be tightened about one to one-and-a-half turns more. See your service manual for exact specifications.

To do a ring job, you will have to remove the oil pan, take off the connecting rod caps, and then take the pistons out of the cylinders. The ridge at the top of the cylinder which is caused by wear from the piston ring will probably have to be removed with a special ridge reamer before the pistons

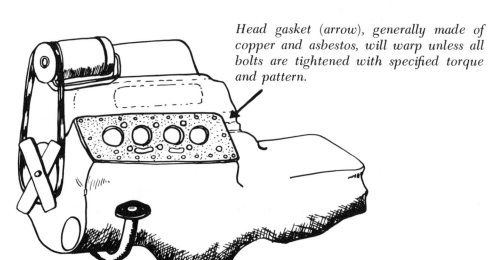

Head gasket (arrow), generally made of copper and asbestos, will warp unless all bolts are tightened with specified torque and pattern.

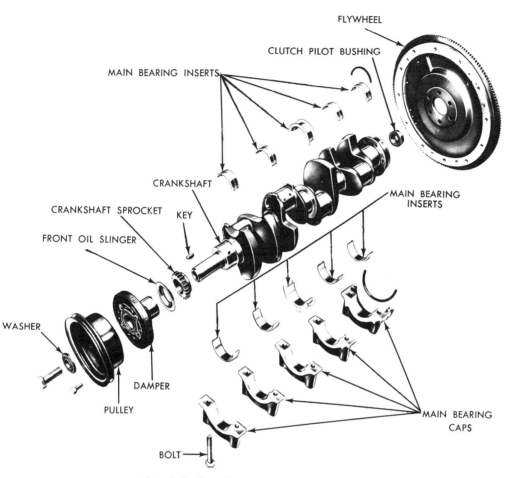

Crankshaft and related parts. (Ford)

Torque pattern specified for most cars will look something like this. Numbers refer to order in which bolts should be tightened. It is always best to check the pattern for a specific car in a service manual.

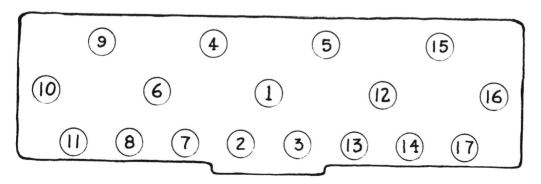

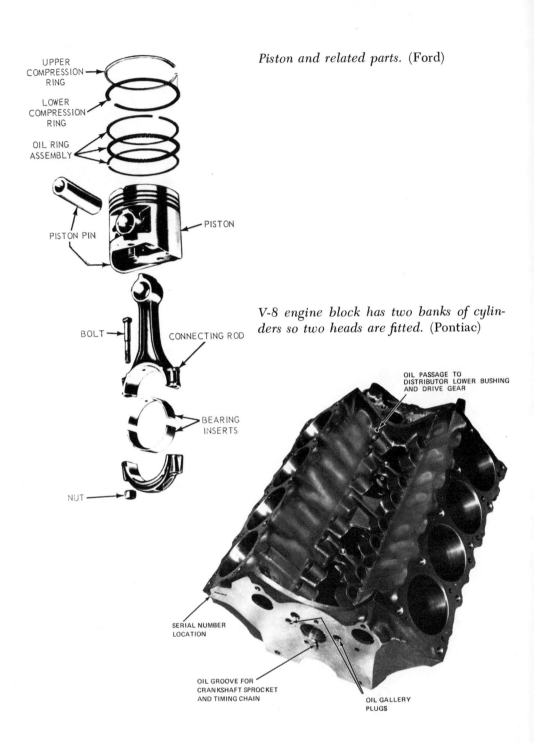

Piston and related parts. (Ford)

UPPER COMPRESSION RING

LOWER COMPRESSION RING

OIL RING ASSEMBLY

PISTON PIN

PISTON

BOLT

CONNECTING ROD

BEARING INSERTS

NUT

V-8 engine block has two banks of cylinders so two heads are fitted. (Pontiac)

OIL PASSAGE TO DISTRIBUTOR LOWER BUSHING AND DRIVE GEAR

SERIAL NUMBER LOCATION

OIL GROOVE FOR CRANKSHAFT SPROCKET AND TIMING CHAIN

OIL GALLERY PLUGS

will slip out of the bore. Then the cylinders must be measured for wear and honed to remove the glaze. The old piston rings must be removed and new ones of the correct size fitted. A ring spreader and compressor are the special tools for this job. The piston skirts will probably need resizing. Finally the pistons must be replaced, after the cylinder bore has been coated with a film of oil, and the connecting rods attached again.

Certainly you will need some professional help in doing a ring job. Unless you can get an engine jobber to come to your home, you would have to remove the engine and transport it back and forth.

The best solution is often to remove the head and pistons yourself at the shop of a cooperative mechanic or engine jobber, or perhaps at a school where auto repair is taught and the proper equipment is available. There are many areas where adult education courses are given at local high schools with well-equipped shops. You can have a jobber work on the pistons or do the job yourself under the supervision of a teacher.

On many cars it is difficult to remove the oil pan because the engine has to be jacked up on its mounts for clearance. Then it will probably require some determined jacking and levering to get everything lined up when you refit the pan. So, while the pan is off, you should consider replacing worn main bearings and having the crankshaft journals checked for scoring and possibly reground.

How much work to do on an engine which is well-worn is always a question. The answer will depend on the availability of parts, and the services of a friendly automotive machinist or auto shop teacher who can help you over the rougher steps.

Grinding valve seats is one of the services of an automotive machine shop.

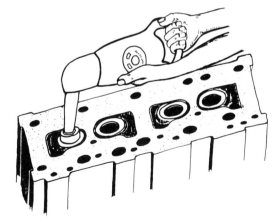

CLUTCH

The transmission or engine must be removed to get at the clutch. On some cars you may only have to disconnect the driveshaft, supporting the weight of the transmission with a jack, and lever the transmission back just far enough for clearance. Then mark the clutch cover to flywheel alignment, and take the clutch cover off after unfastening the bolts. They are loosened one turn at a time and evenly all around to prevent warping the clutch cover.

On other cars the flywheel housing must be removed, or the recommended procedure may call for engine removal. The service manual for your car will give a step-by-step guide.

Once you have gone to the trouble of getting at the clutch, you should always put in a new friction disc, clutch pilot bushing, and throwout bearing. Replacing these parts is an inexpensive precaution which could prevent you from having to do the job over in a year or two. The clutch pressure plate should be changed if it has been in service for an extended period or you have reason to believe it is faulty.

Never get grease or moisture on the clutch disc or flywheel face. The flywheel face and pressure plate can be cleaned with a solvent such as carbon tet and wiped dry. Then the disc and flywheel face should be lightly sanded with a piece of crocus cloth.

Lubrication should be confined to a sparse coat of Lubriplate on the pilot bushing and throwout fork.

The new clutch disc must be installed with the correct side facing the flywheel and is usually marked.

A clutch aligning tool or an old transmission input shaft from a junkyard must be used to line up the clutch disc with the pilot bushing and keep it in line while you refasten the clutch cover. It looks easy to do it by eye, but it really isn't.

It may take a little doing to move the transmission forward and pass the input shaft back through the clutch and pilot. You will need an assistant to do some jacking or levering at the rear of the transmission case as you guide the shaft in. Tighten retaining bolts only after the transmission is fully seated in place.

The final step is to check the clutch pedal free play and adjust the linkage. (See Chapter 5.)

POWER TRAIN

There are such a variety of manual and automatic transmission designs that it is impossible to comment on disassembly or repair in this brief chapter. Generally you will be better off leaving automatic transmission repairs to a garage, although vacuum modulator replacement, adjusting clutch bands, shift linkage adjustment, replacement of a defective governor, and replacement of a rear seal are jobs within the capability of the average amateur mechanic.

Manual transmissions are less complicated mechanically. The special tools required for most repairs are confined to snap-ring pliers, an accurate set of feeler gauges, gear pullers, and punches of various diameters. Most repairs will require that you remove the transmission from the car. The disassembly procedure can involve some hard work, but should not prove too taxing to those who follow the steps in a service manual and work methodically. Broken or chipped gear teeth, worn synchronizers, and improper clearances are readily apparent once the lube has been drained and the grease wiped off. When defective parts are replaced, new snap-rings and thrust washers should be used. New gaskets are always required.

Before attempting any work on a manual transmission, however, always consult a service manual. Sometimes special jigs or other factory-specified tools are required.

The replacement of a U-joint or the installation of U-joint rebuild kit will probably also require some good snap-ring pliers, although some U-joints are held together with U-bolts or wing rollers.

Before disassembling a U-joint and removing the shaft, it is important to mark the part in front of the joint (flange yoke) and behind it (slip yoke), as well as the shaft itself to be sure the parts can be lined up accurately during re-installation.

The shaft should always be handled carefully, supported with a stand when a U-joint is removed, and never crushed in a vise.

There are lots of different arrangements for joints and shafts, with cross and roller U-joints, ball and trunnion U-joints, Hotchkiss or torque-tube drive, two-piece drive shafts with three U-joints and a center support, etc. Removing the drive shaft on some torque-tube drive designs can be a real hassle. But normally servicing U-joints and drive shafts is not all that difficult because they are not very complicated mechanically.

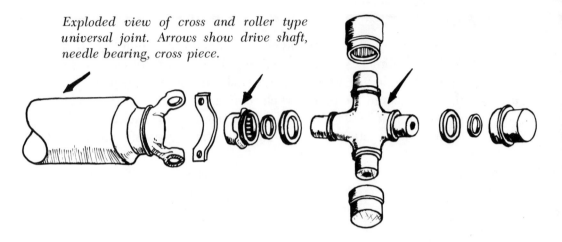

Exploded view of cross and roller type universal joint. Arrows show drive shaft, needle bearing, cross piece.

When the snap-rings or retainers of the U-joint are unfastened, it can be disassembled by placing it in a vise positioned so that each bearing cap can be driven out with a flat-nosed punch. Then the cross assembly can be removed.

On some cars it is necessary to plug the back of the transmission housing when the drive shaft is removed to avoid having transmission fluid run out.

When reassembling the joint with new parts from the rebuild kit, be sure to pack the needle bearings with grease. SAE 140 is usually specified.

In cross and roller U-joints, the rollers are seated in place, with the needle bearings inserted, and driven in until flush, with the jaws of a vise. A soft-faced hammer can be used to strike the yoke and seat the roller against snap-rings.

Some U-joints have a grease fitting for regular lubrication. They should be greased with a hand gun fitted with the proper adapter. It is particularly important to keep constant velocity joints well greased.

The shaft itself requires no routine service and is one of the most trouble free parts of a car. When it does get out of balance, however, the condition can be difficult to correct. Imbalance due to improper shaft installation or an irregular or damaged shaft can be corrected by proper installation or fitting of a new shaft. Vibration caused by an incorrect shaft-to-differential pinion angle, which is usually only a problem on cars where the engine or rear-end has been swapped, will require expert attention.

Minor imbalance problems can sometimes be corrected by attaching two

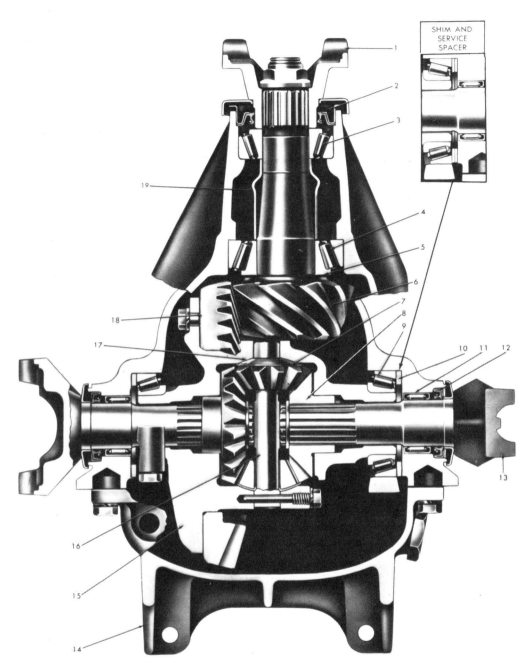

Center section of rear axle. Parts indicated are (1)companion flange; (2)pinion seal; (3)front pinion bearing; (4)rear pinion bearing; (5)pinion shim; (6)pinion; (7)differential pinion; (8)differential side gear; (9)differential bearing; (10)differential bearing shim; (11)yoke bearing; (12)yoke bearing seal; (13)side gear yoke; (14)carrier cover; (15)differential case; (16)differential pinion shaft; (17)thrust washer; (18)ring gear; (19)pinion bearing spacer. (Chevrolet)

A good tool for removing hub and axle shaft is the slide-hammer puller. (Owatonna)

large hose clamps (worm-drive type) around the shaft, and rotating them until vibration disappears.

At the rear end of the power train your primary efforts should probably be confined to pulling the axle shafts or replacing a grease seal. To those accustomed to working on manual transmissions, differential service will not be all that difficult, but it does require some special equipment, and the various differential designs, including the limited-slip variety, present a problem.

Since the differential housing can be readily removed on most cars, you can save part of the cost of repairs by taking out the unit yourself and bringing it in for service.

After the retaining plate is unbolted, a slide hammer puller is used to remove most axle shafts. Those with a tapered end require use of a special wheel hub puller. Shafts with an inner end lock can be freed by removing the differential cover, unbolting the differential pinion shaft, and pushing the axle shaft inward to free it. The C-washer which retains the shaft will drop out, allowing the axle to be withdrawn.

When an axle breaks with part of the stub left in the housing, the broken piece must be fished out with a tool made for that purpose and the housing must be drained and flushed to remove any broken particles.

Outer grease seals can generally be pulled off with an impact puller and driven back on with a flat punch or collar wide enough to ensure that the seal stays squarely seated while being inserted. The lip of the seal should face inward.

Replacing the bearings on an axle shaft is a job best left to an automotive machinist. The charge won't be much.

PAINTING

One of the easiest ways to save money on fixing up your car—and get a superior job to boot—is to paint it yourself.

If you feel diffident about your talents with a spray gun, you can usually make a deal with a local body man to do all the sanding and masking yourself and have him handle the actual painting. Since your part of the job will be the real drudgery, this method should save at least half the cost and probably more.

The first step is to remove the bumpers, door handles, and all the chrome trim you can, and carefully tape over any pieces that are fixed in place. You'll also have to remove or mask off rubber pieces around doors or windows and any rubber bumpers on the doors, hood, or trunk. Old newspapers can be taped in place to mask off the glass when you are ready to begin painting.

Minor dents and even holes can be patched using the procedure covered in Chapter 5. Any extensive work, which requires leading or extensive patching with fiberglas, and the use of welding equipment, is best left to the body shop.

Wet sanding comes next. A water supply from a garden hose is best, but the bucket and sponge technique will do. The sandpaper should be a fairly coarse grit (#220 or so) backed with a rubber sanding block, over large, flat areas, and hand-held when necessary. Steel wool can be used in really tight areas around trim.

Keep the area you are sanding wet at all times. When you come to areas which have been repaired, featheredge them, so underlying coats of paint show up in wide borders. Rust spots can be removed with a disc sander chucked in an electric drill and then gone over by hand.

When you are finished sanding the whole car, go over it again with a finer grit paper (#320) and then wipe it off, clean it with a commercial degreaser, and wash the degreaser off.

If you plan to spray the paint on yourself, wet down the area of your garage where you plan to work, to keep dirt down, and let it dry. Sweeping or vacuuming first is a good idea if the garage is pretty dusty. The more meticulous you are about cleaning, the better your job is likely to turn out. Spread newspapers all around to catch paint overspray.

You can rent a compressor and spray gun from a large paint store that supplies body shops. Here is where you can also get advice on which paint to use (enamel or acrylic lacquer are most popular nowadays) and how many coats to apply. Along with the paint you buy, get primer, thinner, and a tack cloth for going over the body and picking up dust just before you paint.

Mix up your primer with thinner and practice spraying it on some cardboard to see how everything works. You'll want to keep the gun about a foot away from the car while you paint, be sure the pressure is up to the point specified, and adjust the nozzle of the gun so the paint goes on smoothly rather than in bands or blotches. Keep the gun pointed squarely at the surface to be covered, and spray in long, overlapping, sideward strokes, pressing the trigger just after the stroke begins and releasing it before the stroke ends.

Always spray at a temperature above 50° and below too damn hot. Start at the top and then work downward as a general principle.

If the primer runs or sags, there's no harm done since you can always sand again and spray on some more. This is the time to practice until you are confident of your abilities with the spray gun.

The spray gun should be cleaned with thinner after use. Be sure the cover vent holes and nozzle holes are not clogged or the spray pattern will be faulty.

It may be necessary to use a primer sealer if you are painting over colors that "bleed" or painting lacquer over enamel. Follow directions with the primer or ask the paint shop man.

Allow the primer to dry and shrink for at least a week. Your car will look kind of funny if you drive it around, but the eventual result will be worth it.

Then wet-sand with a finer paper (#400 will do), degrease and wash off again, and remove dust with a tack cloth.

The final painting procedure depends upon whether you're using acrylic lacquer or enamel.

Enamel dries relatively slowly. Spray on one light coat, wait a few minutes, and then spray on the next. Three or four coats should do and no further treatment is necessary.

Acrylic should be sprayed on in three or four light coats and you can continue with the next as soon as one coat is done. However acrylic must be wet-sanded with very fine paper (#600) after it has dried for a few days and before it is buffed to a finish. For a really first-rate job, spray a few more coats at that time (using paint diluted with about half again as much thinner as normal) and sand again. The final buffing is done with very thin rubbing compound.

Runs or sags in acrylic or enamel can be repaired (sort of) by wet-sanding lightly with fine grade paper and spraying a thin mist coat over the general area. With enamel, allow the paint to dry for a few days before you try to correct your errors.

An enamel paint job should not be waxed or even washed for a few months after the paint is applied. Just stand around and admire it.

Okay, your car is all shiny again and running perfectly. Now all you have to do is get around to cleaning up that mess in the garage.

11

Wheeling and Dealing

WARNING: *In this chapter the actual costs of owning a car are considered in detail. This is a topic which many car enthusiasts will find obscene. Since the intent of the chapter is to show how to save money on the purchase and operation of a motor vehicle, the redeeming social value should be apparent. Nevertheless, there is much explicit material which some may find offensive.*

● In the days when people were wondering whether to buy a new car or keep their horse and buggy, I am told that the car salesmen had one argument which was a clincher. "A horse keeps on eating hay whether you ride it or not, while the Model T doesn't use a drop of gas unless you drive it," the salesmen would say.

The logic was debatable back then; even more so today. The sad truth is that cars do eat while standing in the garage and the two major constituents of their diets are chunks of protein known as "depreciation" and "insurance."

By tradition the costs of owning a car are determined on a mileage basis. From an economic viewpoint this method may be most valid; psychologically it can be deceptive.

Consider, for instance, the costs of operating an "average" full-sized car, determined to be 13.6 cents per mile in a recent federal study.* The study might be used as a basis for advising the owner of such a car to take a bus to work rather than commuting by motor vehicle. The comparative costs are $1 round-trip for the bus, and, since the distance is ten miles each way, a total of $2.72 by car.

"Nonsense," says the car owner (whom we will call Joe since he's so average). "The major costs of operating a car include depreciation and insurance as well as gasoline. Depreciation and insurance are the same whether I use the car or not. My only additional cost for driving twenty miles is gas

* L. L. Liston *and* C. L. Gauthier, Costs of Operating an Automobile, *U.S. Department of Transportation, 1972.*

TABLE 1. ESTIMATED COST OF OPERATING A STANDARD 1972 MODEL AUTOMOBILE [1]

(Total costs in dollars, costs per mile in cents)

ITEM	FIRST YEAR (14,500 miles)		SECOND YEAR (13,000 miles)		THIRD YEAR (11,500 miles)		FOURTH YEAR (10,000 miles)		FIFTH YEAR (9,900 miles)	
	TOTAL COST	COST PER MILE	TOTAL COST	COST PER MILE	TOTAL COST	COST PER MILE	TOTAL COST	COST PER MILE	TOTAL COST	COST PER MILE
Costs excluding taxes:										
Depreciation	1,226.00	8.46	900.00	6.92	675.00	5.87	500.00	5.00	376.00	3.80
Repairs and maintenance	81.84	.56	115.37	.89	242.65	2.11	296.09	2.96	275.54	2.78
Replacement tires	17.90	.12	16.05	.12	23.72	.21	44.40	.44	43.95	.44
Accessories	3.21	.02	3.08	.02	2.96	.02	2.83	.03	2.82	.03
Gasoline	286.75	1.98	257.16	1.98	227.58	1.98	197.72	1.98	195.83	1.98
Oil	11.25	.08	11.25	.09	12.00	.10	12.00	.12	12.75	.13
Insurance [2]	164.00	1.13	156.00	1.20	156.00	1.36	147.00	1.47	147.00	1.49
Garaging, parking, tolls, etc.	208.36	1.44	199.22	1.53	190.08	1.65	180.94	1.81	180.33	1.82
Total	1,999.31	13.79	1,658.13	12.75	1,529.99	13.30	1,380.98	13.81	1,234.22	12.47
Taxes and fees:										
State:										
Gasoline	74.62	.51	66.92	.52	59.22	.52	51.45	.51	50.96	.52
Registration	30.00	.21	30.00	.23	30.00	.26	30.00	.30	30.00	.30
Titling	177.15	1.22	—	—	—	—	—	—	—	—
Federal:										
Gasoline	42.64	.30	38.24	.30	33.84	.29	29.40	.30	29.12	.30
Oil [3]	.22	—	.22	—	.24	—	.24	—	.26	—
Tires	1.38	.01	1.24	.01	1.82	.02	3.42	.03	3.39	.03
Total taxes	326.01	2.25	136.62	1.06	125.12	1.09	114.51	1.14	113.73	1.15
Total of all costs	2,325.32	16.04	1,794.75	13.81	1,655.11	14.39	1,495.49	14.95	1,347.95	13.62

ITEM	SIXTH YEAR (9,900 miles)		SEVENTH YEAR (9,500 miles)		EIGHTH YEAR (8,500 miles)		NINTH YEAR (7,500 miles)		TENTH YEAR (5,700 miles)		TEN YEAR TOTALS AND AVERAGES (100,000 miles)	
	TOTAL COST	COST PER MILE	TOTAL COST	COST PER MILE	TOTAL COST	COST PER MILE	TOTAL COST	COST PER MILE	TOTAL COST	COST PER MILE	TOTAL COST	COST PER MILE
Costs excluding taxes:												
Depreciation	259.00	2.61	189.00	1.99	121.00	1.42	85.00	1.13	48.00	.84	4,379.00	4.38
Repairs and maintenance	292.54	2.95	397.56	4.19	171.82	2.02	244.33	3.26	29.17	.51	2,146.91	2.14
Replacement tires	45.44	.46	50.69	.53	62.79	.74	52.80	.70	42.11	.74	399.85	.40
Accessories	8.57	.09	8.30	.09	7.65	.09	6.97	.09	5.79	.10	52.18	.05
Gasoline	195.83	1.98	188.03	1.98	168.13	1.98	148.22	1.98	112.71	1.98	1,977.96	1.98
Oil	13.50	.14	13.50	.14	13.50	.16	12.00	.16	6.75	.12	118.50	.12
Insurance [2]	116.00	1.17	116.00	1.22	116.00	1.37	116.00	1.55	116.00	2.04	1,350.00	1.35
Garaging, parking, tolls, etc.	180.33	1.82	177.89	1.87	171.80	2.02	165.71	2.21	154.74	2.71	1,809.40	1.81
Total	1,111.21	11.22	1,140.97	12.01	832.69	9.80	831.03	11.08	515.27	9.04	12,233.80	12.23
Taxes and fees:												
State:												
Gasoline	50.96	.52	48.93	.51	43.75	.52	38.57	.52	29.33	.51	514.71	.51
Registration	30.00	.30	30.00	.32	30.00	.35	30.00	.40	30.00	.53	300.00	.30
Titling	—	—	—	—	—	—	—	—	—	—	177.15	.18
Federal:												
Gasoline	29.12	.29	27.96	.30	25.00	.29	22.04	.29	16.76	.29	294.12	.30
Oil [3]	.27	—	.27	—	.27	—	.24	—	.14	—	2.37	—
Tires	3.50	.04	3.90	.04	4.84	.06	4.07	.06	3.24	.06	30.80	.03
Total taxes	113.85	1.15	111.06	1.17	103.86	1.22	94.92	1.27	79.47	1.39	1,319.15	1.32
Total of all costs	1,225.06	12.37	1,252.03	13.18	936.55	11.02	925.95	12.35	594.74	10.43	13,552.95	13.55

1. This estimate covers the total costs of a fully equipped, medium-priced, standard size, 4-door sedan, purchased for $4,379, operated 100,000 miles over a 10-year period, then scrapped. Baltimore area prices, considered to be in the middle range, were used.

2. Previous editions of this study used insurance rates designated for Baltimore city. The rates shown above are for the Baltimore suburbs, and consequently are less than the rates presented in the previous study. If the Baltimore city rates had been used in this study, the insurance costs would have been higher. (For example, the first year would have been $232).

3. Where costs per mile were computed to be less than 1/20 cent, a dash (—) appears in the column.

plus a minor amount of wear and tear. Since my car uses less than two gallons of gas to travel twenty miles, say about 65 cents worth of gas, and the wear and tear plus oil certainly amounts to less than an additional 35 cents, I'm saving money by taking my car. Not to mention the convenience."

Joe is right. At least he's correct in not being convinced by our cost-per-mile argument. Perhaps we can convince him by looking at the overall costs of operating a car on a daily basis.

A per-mile cost of 13.6 cents for a car driven an average of 10,000 miles per year would give a daily cost of $3.73. Per month (thirty days), the cost would be $111.90.

A few further computations using the basic data in the Department of Transportation study which is summarized in the tables may help to convince Joe that his car expenses are a major part of the family budget. Before we go further, however, a few of the assumptions made in this study should be noted.

The figures are based upon operating a car for ten years and 100,000 miles. While very few people will operate a car this long, the authors feel that it is not unreasonable to assume that this is the total life of a car as it passes through two, three, or even more owners. They state: "Vehicle survival data developed on the popular size, popular brand cars show that half of the autos were still on the road at the end of ten years."

Another assumption to note is that mileage driven per year diminishes as the car gets older. While the average for ten years is 10,000 miles per year, the actual mileage assumed in the study is 14,500 miles the first year, decreasing to 5,700 miles by the tenth year. By this time, according to the authors, the car is likely to have become the second or third car in a family since it is probably no longer reliable for long trips or extensive daily commuting.

Since the average period of car ownership for those who buy a new car is somewhere around thirty months, it is quite reasonable to assume that Joe trades his car in every three years. If he operates his car according to the mileage figures in the table covering standard-sized vehicles, he will have gone 39,000 miles at trade-in time. His costs are now up to $4.74 per day or $142.20 per month. That's a lot.

The cost is even greater if we consider one major factor which is deliberately excluded from the DOT study. The authors say:

Automobile financing charges are not included in the costs shown in this report. Even if an automobile owner buys his car outright, he is foregoing the opportunity to earn interest on the money used for purchase. Complete cost accounting would require that financing costs or an interest lost charge be included as a part of the automobile costs. At the conservative rate of 6 per cent this would add about $260, or nearly 2 cents per mile during the first year of the life of the standard size car.

According to a recent survey, only 34 per cent of new car buyers paid cash for their cars while the remaining 66 per cent financed all or part of the cost.* We can reasonably assume that Joe is in the latter category. We'll make a further assumption and say that Joe is trading in exactly the same kind of car he is buying, except the former is three years older. Now we can figure the trade-in from the table (it would be $1,578) and find that he has to finance $2,801 of his new car cost. Dropping the odd dollar and giving Joe a $300 discount, which is about what he can expect on such a purchase, we find that Joe must finance $2,500 of the purchase price.

Joe borrows the money at an 8 per cent yearly rate (not a true 8 per cent, but a nominal one on a discounted loan—detailed figures on this transaction are given further on in this chapter in the section on finance). He finances the car over a three-year period. Again we look at Joe's total costs for the three years, subtracting a $300 discount and adding financing charges of $789.45. His total costs for the three-year period are now $5,-675.88, daily costs are $5.18, and the monthly tab is $155.46. The costs are escalating and we're not done yet.

How many cars are there in Joe's household? According to figures from the 1970 census, there are 24.6 per cent of American households with two cars and 4.7 per cent with three or more cars. In the middle-income groups there are 29.9 per cent of households with an income between $7,500 and $9,999 which have two cars, 40.2 per cent with an income of $10,000 to $14,999 which maintain two cars, and 50.4 per cent earning more than $15,000 and having two cars. These income levels represent total household income contributed by all working members of the family.

Joe is a moderate earner. Even with his wife's earnings from a part-time job, the family income is only $12,500. In this income bracket there is a

* In 1969. The survey was carried out by the University of Michigan Survey Research Center and figures here are taken from the Automobile Manufacturers Association publication, 1971 Automobile Facts and Figures, p. 44.

TABLE 2. ESTIMATED COST OF OPERATING A COMPACT SIZE 1972 MODEL AUTOMOBILE [1]

(Total costs in dollars, costs per mile in cents)

ITEM	FIRST YEAR (14,500 miles)		SECOND YEAR (13,000 miles)		THIRD YEAR (11,500 miles)		FOURTH YEAR (10,000 miles)		FIFTH YEAR (9,900 miles)	
	TOTAL COST	COST PER MILE	TOTAL COST	COST PER MILE	TOTAL COST	COST PER MILE	TOTAL COST	COST PER MILE	TOTAL COST	COST PER MILE
Costs excluding taxes:										
Depreciation	674.00	4.65	519.00	3.99	394.00	3.42	305.00	3.05	243.00	2.46
Repairs and maintenance	79.41	.55	107.14	.83	170.61	1.48	218.90	2.19	240.27	2.43
Replacement tires	15.30	.11	13.71	.11	12.13	.11	34.27	.34	33.93	.34
Accessories	3.21	.02	3.08	.02	2.96	.03	2.83	.03	2.82	.03
Gasoline	244.25	1.68	218.97	1.69	193.68	1.69	168.39	1.68	166.78	1.68
Oil	10.50	.07	10.50	.08	11.25	.10	11.25	.11	12.75	.13
Insurance	155.00	1.07	147.00	1.13	147.00	1.28	140.00	1.40	140.00	1.41
Garaging, parking, tolls, etc.	208.36	1.44	199.22	1.53	190.08	1.65	180.94	1.81	180.33	1.82
Total	1,390.03	9.59	1,218.62	9.38	1,121.71	9.76	1,061.58	10.61	1,019.88	10.30
Taxes and fees:										
State:										
Gasoline	63.56	.44	56.98	.44	50.40	.44	43.82	.44	43.40	.44
Registration	20.00	.14	20.00	.15	20.00	.17	20.00	.20	20.00	.20
Titling	109.86	.75	—	—	—	—	—	—	—	—
Federal:										
Gasoline	36.32	.25	32.56	.25	28.80	.25	25.04	.25	24.80	.25
Oil [2]	.21	—	.21	—	.22	—	.22	—	.26	—
Tires	1.17	.01	1.05	.01	.92	.01	2.61	.03	2.59	.03
Total taxes	231.12	1.59	110.80	.85	100.34	.87	91.69	.92	91.05	.92
Total of all costs	1,621.15	11.18	1,329.42	10.23	1,222.05	10.63	1,153.27	11.53	1,110.93	11.22

ITEM	SIXTH YEAR (9,900 miles)		SEVENTH YEAR (9,500 miles)		EIGHTH YEAR (8,500 miles)		NINTH YEAR (7,500 miles)		TENTH YEAR (5,700 miles)		TEN YEAR TOTALS AND AVERAGES (100,000 miles)	
	TOTAL COST	COST PER MILE	TOTAL COST	COST PER MILE	TOTAL COST	COST PER MILE	TOTAL COST	COST PER MILE	TOTAL COST	COST PER MILE	TOTAL COST	COST PER MILE
Costs excluding taxes:												
Depreciation	194.00	1.96	152.00	1.60	103.00	1.21	73.00	.97	39.00	.68	2,696.00	2.70
Repairs and maintenance	268.81	2.72	412.04	4.34	177.27	2.09	78.95	1.05	31.10	.55	1,784.50	1.79
Replacement tires	38.45	.39	36.89	.39	61.53	.72	54.29	.73	41.27	.72	341.77	.34
Accessories	8.57	.09	8.30	.09	7.65	.09	6.97	.09	5.79	.10	52.18	.05
Gasoline	166.78	1.68	160.06	1.69	143.11	1.69	126.43	1.69	96.03	1.68	1,684.48	1.68
Oil	12.75	.13	12.75	.13	12.75	.15	12.00	.16	6.75	.12	113.25	.11
Insurance	114.00	1.15	114.00	1.20	114.00	1.34	114.00	1.52	114.00	2.00	1,299.00	1.30
Garaging, parking, tolls, etc.	180.33	1.82	177.89	1.87	171.80	2.02	165.71	2.21	154.74	2.72	1,809.40	1.81
Total	983.69	9.94	1,073.93	11.31	791.11	9.31	631.35	8.42	488.68	8.57	9,780.58	9.78
Taxes and fees:												
State:												
Gasoline	43.40	.44	41.65	.44	37.24	.44	32.90	.44	24.99	.44	438.34	.44
Registration	20.00	.20	20.00	.21	20.00	.23	20.00	.26	20.00	.35	200.00	.20
Titling	—	—	—	—	—	—	—	—	—	—	109.86	.11
Federal:												
Gasoline	24.80	.25	23.80	.25	21.28	.25	18.80	.25	14.28	.25	250.48	.25
Oil [2]	.26	—	.26	—	.26	—	.24	—	.13	—	2.27	—
Tires	2.93	.03	2.81	.03	4.69	.06	4.15	.06	3.15	.06	26.07	.03
Total taxes	91.39	.92	88.52	.93	83.47	.98	76.09	1.01	62.55	1.10	1,027.02	1.03
Total of all costs	1,075.08	10.86	1,162.45	12.24	874.58	10.29	707.44	9.43	551.23	9.67	10,807.60	10.81

1. This estimate covers the total costs of a medium-priced, compact size, 2-door sedan, purchased for $2,696, operated 100,000 miles over a 10-year period, then scrapped. Baltimore area prices, considered to be in the middle range, were used.
2. Where costs per mile were computed to be less than 1/20 cent, a dash(—) appears in the column.

40.2 per cent chance that there are two cars in the family and an 8.2 per cent chance that they operate three or more cars. In addition, he lives in the western part of the United States, where multiple car ownership is slightly higher, and he is between thirty-four and forty-four years old— another fact increasing the likelihood that the family operates two or more cars. Chances are better than even that Joe and his wife have at least two cars.

Since Joe is such an average type, his second car should not cost as much as the first. Let's assume it is a compact, purchased used after the third year of service and kept through the seventh year before trading. Looking at Table 2 we see that Joe's total costs (fourth through sixth year for a compact car) are $2,065.15. The purchase price of the car is $1,109, a figure low enough so Joe could pay cash for the car. Again we say that Joe's previous car was exactly the same type and (from the table) we credit him with a trade-in of $357. His interest-lost figures on the $742 he paid in cash for the car are added to the total costs, giving a new total of $2,198.71. His daily costs are $2 and the monthly tab is $60.

To operate both cars, Joe is paying $7.18 per day and $215.40 each month. Since we said his income was $12,500 per year, he is spending just over 20 per cent of this income on cars.

We can now offer Joe a more convincing picture of what it is costing him to run two cars. Each week when he gets his pay check of about $170 (after taxes), he can deduct about one-fifth of his gross pay, or a little over $48, for car expenses. Whether he can afford it or not is a question only Joe can answer.

WHAT KIND OF CAR?   WHICH OPTIONS?

Should we convince Joe to buy a cheaper car? An older car? One with fewer accessories? Or will he save money by selecting accessories which will increase the car's resale value. Possibly he might even consider leasing a car.

Everyone has his own version of sensible car economies. Here I will give you a rundown on mine, attempting to be as objective as possible. Perhaps the best way to begin is by looking at a few persistent myths.

For instance, there's the story about the car which is actually appreciating in value, such as the twenty-year-old MG which is now worth more than it sold for brand new.

The deceptive part of the story is that the costs over the intervening twenty years have not been mentioned. Nobody considers all the sweat, tears, and parts-hunting required to maintain a car in good condition for that length of time. Also the chances are that any classic sports car is used only as a second or third car, driven on sunny days, and always garaged (so the cost of garage rental or maintenance must be added to the total).

And what if the MG were in an accident and totally wrecked? The insurance company would probably pay the owner just enough to cover the cost of a new set of tires and wheels.

Despite these caveats, resale value should play a part in the choice of a car, although not as large a part as some would like to think. In general, it is best to avoid extreme cars, cars bedecked with stripes, scoops, and other dress-up items which may go out of style, and the types of vehicles (dune buggies, for example) that are a passing fad.

You can check *Consumers Reports* or any of the used-car price guides to determine the well-established makes and models which consistently have a better-than-average resale value.

If you intend to trade in a car after only a few years of ownership, it is worthwhile considering how its resale value will be affected by the options you select.

The basic options which will return a fair portion of their cost at resale time are an automatic transmission, air-conditioning, power brakes and steering, and a radio. In order to have a significant trade-in value, however, these options must be consistent with the character of the car.

A luxury sedan without an automatic transmission would be difficult to sell. Conversely, a small foreign economy car might be hard to sell *with* an automatic transmission since buyers are looking for economy and minimal maintenance. In the same way, on high-performance cars the luxury options are secondary to items such as four-speed manual transmissions, bucket seats, and wide tires.

Back to our friend Joe. He may decide to avoid high depreciation costs by purchasing a used car. Naturally he would like to buy a used car which is "like new," little realizing that this is another myth to delude and ensnare the innocent car buyer. Supposedly such a car was owned by an eccentric millionaire who hardly drove it out of the showroom before he decided that the paint color didn't suit him. Or it may be an older car which belonged to the proverbial little old lady who only took it to church on Sundays.

No matter how the story is told, it still cannot alter the fact that there is a wide gulf between any new car and one which is even slightly used. The degree of risk is greater, the buyer can no longer choose the options he prefers, and a portion of the mileage which is anticipated to be trouble free has been skimmed off the top like the cream in a bottle of milk.

If the "like new" used car was really just driven around the block before it was traded in, the dealer would probably sell it to his brother-in-law. And the little old lady who drove only as far as church on Sundays probably never heard of a grease job.

The total cost of running a used car *is* usually less than that of a new one. And the older used car can provide serviceable transportation in Joe's family, while a carefully selected vehicle of more recent vintage may very well be a reliable first car to take Joe to work and back every day. However the "like new" car which is a fantastic bargain just doesn't exist.

Whether Joe buys a new or moderately used car, the type of vehicle he selects can play a big part in determining his ultimate costs.

Its size should be somewhere between that of a Volkswagen and the size of an American intermediate model. The real minicars are just not suited to turnpike cruising and hauling chores, while the largest American cars have little, if any, more interior space than the one-size-smaller intermediates.

Minicars will probably become more popular and useful as strictly urban cars are mandated. By this time, the full-sized car—which is an overgrown giant by foreign standards—will probably be headed toward oblivion.

A four- or six-cylinder engine offers adequate power on an intermediate car or a compact and is less expensive to buy and maintain.

Sedans are lower priced than the more glamorous hardtops and convertibles, and sedans or wagons have more interior room.

For those who can really utilize their special features, vans or pickup trucks may be a good choice. While small trucks lack some of the comfort and safety features standard on family sedans, they do have extra load space, superior rough-road capabilities, and better-than-average resale value due to their conservative styling and sturdy construction.

Utility vehicles have become popular for camping and off-road travel. They can be a way to have fun and save money to boot, so long as they are selected with an eye toward the practicalities of the situation.

There are extreme cases of people who buy ten-thousand-dollar motor-homes in order to save motel bills on a two-week vacation once a year, or those who buy a four-wheel-drive truck with every utility and heavy-duty option (at about the price of a new Cadillac) and use it only for fetching groceries from the corner supermarket.

At the opposite extreme are those optimists who load heavy campers on little half-ton pickups or try to haul eight-thousand-pound trailers with three-thousand-pound cars. For safety's sake, and to keep the family budget healthy, it pays to gain some experience by renting recreational vehicles before the purchase of one is considered. It's also a good idea to consult manufacturers and experienced dealers about tow car options, hitches, and maximum safe weights on pickups or motorhomes.

Foreign cars have penetrated the American market to the extent where they offer a considerable challenge to Detroit. Those in the market for an economy car, a moderate-sized luxury sedan, or a sports car will find many foreign makes to choose from.

The advantages of foreign cars include low price and good gas mileage (on economy models), construction quality and detailing generally superior to the Detroit standard (especially on German and Japanese cars), and agile handling. Many makes offer more comfortable seats, better brakes, and a more complete line of standard features than their American counterparts.

Detroit still leads in the design of some components such as automatic transmissions and air-conditioning. American cars have more power for effortless high-speed cruising, a complete selection of optional equipment for better performance and luxury, and their durability is frequently superior to that of the average foreign car with a heavily stressed engine.

Parts and service availability is another area where Detroit wins out. While some of the more popular makes such as Volkswagen, Datsun, and Mercedes maintain dealer networks with a good reputation, they can't match Chevy, Ford, or Plymouth when it comes to total number of dealers, availability of repairs at independent garages, and the amount of parts in stock at the run-of-the-mill supply house.

Joe's need of reliable, low-cost transportation should lead him to buy a moderate-sized sedan or wagon. A foreign car can be a good choice if there is a nearby dealer for one of the more popular makes. In the event that Joe chooses to buy a used car, it should be an easily serviced model which is not more than three years old and shows no signs of excessive mileage.

It is hard to specify exactly how long Joe should keep his car. All we can say is that, when the car reaches the sixty- to eighty-thousand-mile mark or is more than five years old, it will no longer pay Joe to make any major repairs on it and expect it to continue functioning reliably as his primary transportation. When this point is reached, Joe can either trade in his car or keep it for use as a second car.

Since Joe is average in every way, the alternative of leasing a car is not for him. A long-term lease offers advantages primarily to those who put a great deal of mileage on their cars and, for income tax purposes, can deduct the leasing cost as a legitimate business expense. The short-term lease is advantageous for those who need a car only part of each year. Joe does not fit either of these categories.

Assuming that Joe decides to purchase a new car and keep it as long as he can, the options he selects should be based more upon convenience than anticipated resale value. He should look for those options which add to the safety, reliability, or comfort of the car and avoid those which offer no substantial advantage and may lead to servicing problems.

If Joe lives in a colder part of the country, a limited-slip differential will be an asset on ice and snow while a rear window defroster will aid in maintaining bad weather visibility.

Disc brakes, inertia reel seat belts, and quality tires should be high on his list of desired options. Heavy-duty shocks are also worth the price, since they will last longer than the puny original equipment ones fitted to many cars, and maintain better control.

Other heavy-duty items—suspension, cooling system, battery, alternator —should be fitted to cars that will be subjected to severe use. They are especially necessary on a car that will be used to tow a heavy trailer.

Other useful options that Joe would do well to consider are dashboard gauges to replace idiot lights, bumper guards, outside mirrors, and additional fog or driving lights. Since antiquated state laws limit manufacturers to sealed-beam headlights, Joe will have to see his local auto accessory store if he wants a set of the more desirable quartz halogen lights.

Along with this enumeration of desirable options (the hit list) there's that other list of options to avoid. Pop-open headlights, recessed windshield wipers, automatic speed controls, power door lock systems, and vinyl roof covers are at the top of my list in this category. They all seem to offer very little for the price and to be potential maintenance problems. There are other little gadgets and decor items, such as automatic headlight dimmers, tape stripes, and tires with raised white lettering, which deserve a place on this list.

Above all, our advice to Joe is to be sensible and to choose an inexpensive, moderate-sized car which is likely to offer long and reliable service rather than flash and dash.

Whether Joe will listen to this advice is another story. He's probably got a whole long list of reasons why the most sensible car for him is a new Panther Super Sports with a 483-cubic-inch engine, the Le Mans styling package, and the Custom Deluxe interior which his wife admires.

We'd better not argue with Joe either. Otherwise he is likely to ask how come we're not driving that sensible six-cylinder compact sedan that we recommend. And that's an argument that hits too close to home.

WHERE AND WHEN TO BUY

Just as there are no surefire rules for making a million on the stock market or bringing up your children to become pillars of the community, there are no infallible "how-to's" for coming out a winner in the car-buying game. All I can offer is the sadder-but-wiser advice of a sometime loser.

Try to choose a dealer who is close to your home base. Although a dealer who is fifty miles away may sell the same car for a few dollars less, the savings will disappear when you have to make a hundred-mile round-trip each time you get the car serviced under warranty.

Check a dealer's reputation carefully before you do business with him. Some dealerships are oriented toward the fast turnover of new cars at a big discount. In order to increase their profit margin, they will skimp on car preparation and warranty service. This type of dealer also relies on high-pressure salesmanship and tricky financing plans to keep the profits rolling in.

Occasionally the local Better Business Bureau will prove a good source of information on car dealers in the area. However the B.B.B. is primarily interested in investigating complaints rather than acting as a consumer information bureau.

Your friends and acquaintances may turn out to be the best sources of information, especially if they are mechanically knowledgeable types. Mechanics at independent garages can also give valuable advice.

Do a little detective work on your own by stepping out of the showroom at a car dealership and nosing around the service area.

A dealer who cares about service will keep mechanics on duty after normal business hours for the customer's convenience. There will be service hours at least one evening a week and perhaps on Saturdays.

The parts department will be manned by at least one experienced man who can look up the specifications on any part in stock. Some parts departments even sell tools and service manuals to do-it-yourself customers.

On the floor, the service manager will be actively scheduling repairs. He should have a check sheet on which he marks the requested service operations for each car. If he takes the time to discuss the customer's problems and explain how and when repairs will be made, he is definitely a superior sort.

Mechanics should be working steadily in a well-organized area with the essential tools close at hand. They certainly should not punch in a job on the time clock and proceed to take an extended coffee break on time that will be charged for labor.

While no busy service department can resemble a sterilized laboratory, a good one will exhibit a high degree of efficiency beneath the outward dirt and disorder.

ESTABLISHING A PRICE

All new cars come with a sticker in the window giving the manufacturer's suggested list price for the basic vehicle plus each option fitted. This is known as a "Monroney Sticker" after the gentleman who sponsored the Senate bill which makes such price information mandatory.

While the Monroney Sticker gives the suggested list price, everyone knows that it's not the real price because discounting is almost universal.

The question is how to figure out what's a reasonable discount on the car you intend to buy? Trying to do this by estimating the dealer's mark-up can be a complex procedure.

The mark-up on new cars is on a sliding scale. It ranges from about 14 per cent on compacts to as much as 25 per cent on luxury cars. On foreign car sales the pattern is not so clear since it is influenced by both the distribution set-up and country of origin. There is a still higher profit margin on the more expensive foreign cars, but the percentages vary.

The mark-up on accessories is relatively high. The exact percentage depends upon the manufacturer and the particular option being considered. Perhaps the best way of determining how much the dealer pays for a particular car plus accessories is to consult one of the many car price guides on the market. One can be found on most newsstands.

The trouble is that not even the most detailed car price guide can list the dealer's true cost. This is because car manufacturers sweeten the pot by giving the dealer an end-of-year rebate which is a percentage of his car sales for that year. In addition, there may be monetary incentives offered to move a slow-selling model, increase car sales over last year, etc.

There are many more factors which play a part in establishing the discount a dealer is willing to give. For instance, the salesman's commission, the dealer's overhead, his tax situation, the profit he may make on arranging to finance a car, his anticipated profit on servicing the car, his profit (or loss) on dealer preparation and warranty service, and the interest he must pay on unsold cars in stock. There is also the old economic fact of supply and demand, so that, on a newly introduced model in short supply, there may be no discount at all. (This is usual only for a very few cars, mostly foreign luxury and sports models, but it does happen.)

Unless you are both a cost accountant and a mind reader, it's much easier to forget about figuring what a dealer's mark-up should be, and confine your efforts to comparison shopping. Get a quote from at least two and preferably three or four dealers. Specify the same trade-in and financing plan each time, so the quotes are exactly comparable.

It may also pay you to get a quote from United Auto Brokers, a company which sells new cars for about $125 over the dealer's cost in any area of the country.

What's the catch? There really isn't any. On some luxury cars, imports, and trucks the UAB mark-up is more than $125, but it should still be substantially below your local dealer's price.

While I personally have not dealt with United Auto Brokers, they are a reputable firm and can point to many satisfied customers. The procedure for buying a car through them is to send a check for $7.50 (to United Auto Brokers, Inc., 1603 Bushwick Ave., Brooklyn, N.Y. 11207) and ask for a quote on the car you want. You must specify the make, model, and all options desired.

UAB, through its Car/Puter division, will send back a print-out giving both the dealer cost and sticker price for the basic vehicle and each option. The print-out will include the UAB mark-up ($125 over dealer's cost for most cars) and can be used as a direct purchase order.

There are UAB participating dealers in many areas of the country, so that there should be no delivery problem if you live near a major metropolitan area. Otherwise you can still take delivery through your local dealer, but you will be required to pay an extra fee ($50–$75) for this service. Naturally you will also pay for transportation and dealer preparation.

Theoretically your dealer (or any dealer for the make of car you buy) must provide warranty service on request. Actually, having lost the profit on the sale, he may be less than cooperative.

On a basic compact car, your local dealer's discount price may represent a mark-up of $200 or less. In this case the UAB price of $125 over cost (plus the extra delivery fee, if applicable) would probably not be worth the trouble. On more expensive cars, where the mark-up can run as high as $2,500, the UAB price could represent a substantial savings.

After checking prices at two or more dealers and getting a UAB quote, you may still decide to buy from a nearby dealer with a good reputation. At least you will know how much you are paying for the extra service and other benefits you anticipate.

When you are buying a used car, exact comparison shopping is not possible. You will have to rely on a car price guide to establish the approximate worth of a car you are buying or trading in. There is the "Blue Book" (Kelley Blue Book Auto Market Report), the "Red Book" (published by National Market Reports, Inc.), and the "NADA Book" (issued by the National Auto Dealers Association).

These three publications are periodicals and they are put out in regional editions to reflect used-car prices in different parts of the country. There are also versions which cover specialized vehicles like trucks and motor-homes.

If you can't get a friendly dealer to show you his copy, one or another of these publications might be on hand at your local library. Also banks frequently subscribe (to help them determine car loans), so you could ask the guy at the bank if you can look at his copy.

Of course a car in excellent shape will be worth a little more than book value while a junker may be worth much less. A series of tests to determine the condition of a car are given later on in this chapter.

As a general rule, you will be better off buying a used car from a new-car dealer rather than from a dealer who sells only used models. Dealing in used cars exclusively is a rather marginal business. Those who make the best profits usually buy most of their cars en masse at auctions and give them an extensive cosmetic job before they are put on the lot. You are more apt to find a disguised fleet car or former taxi on the lot of a used-car dealer than you are in the used-car department of a franchised dealer.

BARGAINS

This leads us to an old maxim about car buying, which says: "When you find a bargain that's too good to believe, don't believe it."

There are so many gimmicks to catch the unwary that I can only cite a few representative examples.

New-car dealers advertise seemingly low prices on "lo-mileage executive cars" or demonstrators. These cars may really have been demonstrators or lease cars, or they could be high-mileage fleet cars with the speedometer turned back.

End-of-year sales of leftover cars can offer legitimate savings to those who are willing to buy a car off the floor with a limited choice of models and options. However some dealers actually stock up on "leftover" cars because they know that bargain hunters will snap them up—even if the discount is no greater than normal. With some people the illusion of a bargain is all that counts.

There's another drawback to buying a leftover model which should be mentioned. At trade-in time it will be one year older than normal. Unless you intend to keep the car so long that the trade-in price will be academic, you may lose at one end what you gained at the other.

Some dealers specialize in financing come-ons. They advertise no down-payment and extended financing periods. To buy a car with no down-payment, you will have to give some sort of security (such as your household furniture) and the financing charges will be high. If you fail to make the payments, a loan company will not only repossess your car, but also haul away the living room sofa to boot.

The real sharpie dealers combine their favorite gimmicks in a systematic approach which is almost guaranteed to hook the gullible. They advertise fantastic bargains to draw people into their showrooms. This is the first step.

The potential customers find that the bargains are all sold or else they are not really as depicted. However the apologetic salesman is ready to make a deal on another car which is really very good. To show his good faith and anxiety to please, he reduces the price still further and the bargain is consummated. The customer signs a contract and puts down his deposit.

However the salesman says that the deal must be given final approval by the sales manager. This latter individual (known as the "take-over man" in the trade) then comes on the scene with his swift pencil and well-developed hard-sell techniques. He cannot approve the salesman's offer—it was all a mistake—but he is ready to work another deal. And so it goes.

The whole process is a bit like the old fraternity drinking song about the party where there will be only one bar (chorus: "Boooh"), but it will be a mile long ("Yay"), and there will be only three dancing girls ("Boooh"), and only two dresses ("Yay"), for each girl ("Boooh"), made out of cellophane ("Yay"), and so on ad infinitum. The sales team pulls so many switcheroos that the customer is willing to accept anything out of sheer confusion.

Used-car dealers have their own tricks of the trade. One is the "50/50 guarantee" where the dealer will pay half the cost of any necessary repairs for, say, ninety days. Since the dealer also does the repairs, his 50 per cent might well shrink to zero while yours increases to 100 per cent.

A somewhat more subtle ploy is to offer an exchange privilege. If the car you buy proves unsatisfactory, you can bring it back within ninety days and exchange it for any car on the lot with a similar price. The catch is that the dealer offers discounts on all his cars—until you go back to make an exchange and find that all the cars on the lot are suddenly valued at the full price chalked on the windshield.

The list of gimmicks could be extended on and on and you will never be able to anticipate them all. There's just no way for you to win at the game; all you can do is refuse to play. Stay away from the quick-buck dealers and never, never bring your checkbook along when you go out shopping for a car.

WARRANTIES

Dealer preparation and warranty service are the final steps in the new-car cycle.

As you may know, the warranties which go with a new car are carefully designed to define and limit the manufacturer's responsibility. They are a lot less comprehensive than they may seem at first glance.

It is up to the dealer to determine which repairs fall under the warranty and which are due to abuse, misuse, or accident. Parts which normally require periodic replacement—spark plugs, ignition parts, brake linings—are not covered at all. Tires and batteries may be covered by a separate warranty issued by the company supplying them. This can be inconvenient.

Most warranties do not cover miscellaneous expenses which may be associated with a repair (car rental, for instance).

Although the warranty on a new car tends to leave you at the dealer's mercy, there are a few steps you can take to increase your chances of getting a fair shake.

When you are purchasing the car, get the dealer to agree to give you a signed copy of the pre-delivery inspection check list covering all services that the manufacturer has designated a part of dealer preparation. You will want all items marked o.k. before you take delivery of the car.

Inspect and road test the car as soon as you get it. If you find a major defect, the dealer will probably take it back. Otherwise you can try dumping it on his doorstep and stopping payment on your check. (It may be wise to seek legal advice before taking this step.)

Take the car back to the dealer promptly when you find defects during the warranty period. Call up beforehand to make an appointment and expect the work to be done when promised. Be courteous, but firm.

Keep a record of all repairs carried out and any charges. Make a note if you fail to get service when you want it. Get receipts for any emergency roadside repairs and towing.

Should you fail to get satisfaction from the dealer, despite your best efforts, write to the manufacturer's zone representative in your area. His address should be in your owner's manual. State the problem as clearly as you can and give details of your efforts to get proper service. To aid in clarity, give a brief chronology of problems and the dates when they occurred. Keep the tone of your letter businesslike and don't make any threats. Be sure you save carbons of all correspondence.

As consumer groups have become more vocal, both dealers and manufacturers are accepting their responsibilities more readily and finding that good service has an economic payoff. Your chances of getting satisfaction under a warranty are still far from 100 per cent, but they are getting better.

Be realistic in your expectations. You can't expect a dealer to repaint a car because of a tiny scratch. A bit of hesitation from an engine when you start up in the morning is part of the price you pay for emission controls and there isn't much that the dealer can do about it. On the other hand, when there is a definite malfunction, don't accept a dealer's explanation that "they all do that." Ask *why* they all do that and whether there is a factory service bulletin giving a recommended fix. Write or call the zone representative for clarification, if necessary.

And what do you do when you have a legitimate complaint and nothing seems to work? You kick up a fuss, that's what you do. Be reasonable, but persistent as hell. You can write to the Better Business Bureau, or your congressman, or complain to Ralph Nader. Whatever you do, be sure you have a leg to stand on first and that you have exhausted all the usual channels of complaint. Then prepare for battle and keep at it until you get results.

BUYING AT THE FACTORY

Does it pay to travel to Detroit to pick up your new car? Not unless you're going there anyway to visit Aunt Minnie. Car delivery charges from the Detroit area to the farthest point in the U.S. (with the exception of Alaska

349 · *Wheeling and Dealing*

and Hawaii) will not run much over $225. You can't beat that price by picking it up yourself.

Car buying abroad is another story. If you are in the market for a luxury car, such as a Rolls, Mercedes, or Lamborghini, substantial savings can be realized by picking it up at the factory. Of course you will have to order the car well in advance (a month to six weeks minimum) and be sure it is manufactured to meet U.S. emission and safety regulations.

The savings possible on an economy car are much less. By the time you get through paying the shipping, preparation, port handling charges, and documentation, not to mention the little bribes you have to give to the guys down at the docks to treat your baby with tender, loving care and avoid crunching the fenders as they unload it, your savings will probably be nil.

There is one exception. If you want to tour Europe for a month or more, you can save money by purchasing a car abroad and bringing it back with you. The total cost will be less than renting a car for a long period in Europe plus buying a European car from an importer in the United States.

You can arrange the deal through many foreign-car distributors by specifying delivery in Europe. They can also arrange for insurance coverage during your trip. An international driver's license may be purchased through any branch of the A.A.A.

If your local foreign-car dealer can't make the arrangements, you can contact a firm which specializes in this service. Both Hertz and Avis will do the deed. A few others are Auto-Europe, Inc. (1270 Second Ave., N.Y.C. 10021), European Auto Travel Center (323 Geary St., San Francisco 94102), and General American Shippers (450 7th Ave., N.Y.C. 10001).

For a premium you can make arrangements without the usual four- to six-week waiting period and pick up the car anywhere in Europe. Be sure you are quoted the *full* costs, however, since some dealers mention only the basic shipping charge and leave you to take care of the port arrangements on your own.

The same type of deal is *not* available on Japanese cars. The reason is that Japanese vehicles equipped to meet U.S. regulations cannot be driven in Japan, and those meeting Japanese specifications can't be brought back to the U.S.

FINANCING

Our hypothetical friend Joe had to finance $2,500 of the purchase price of his new car. The dealer told Joe that the finance rate was 8 per cent a year. Then Joe went home to figure how much that was in monthly payments.

Mathematics is not Joe's strong point, although the problem didn't seem too tough. "Let's see," said Joe, "8 per cent of $2,500 (or 2,500 times .08) is $200. Multiplied by 3, since I'm financing the car over three years, it comes to $600. Added to $2,500—that's $3,100—and divided into thirty-six payments. I'll have to pay $86.11 per month."

To be sure his figures were right, Joe consulted his accountant friend Stanley. After glancing at the figures Joe showed him, Stanley only smiled.

"You may be figuring the monthly payments correctly," Stanley said, "but 8 per cent a year isn't your true interest rate. If it was, you'd be paying a lot less than $86.11 a month.

"Look at it this way," Stanley continued. "After your first monthly payment of $86.11 is made, you only owe the bank $2413.89, yet you're still paying interest on the full $2,500. By the end of the second year, after twenty-four payments, you will have paid back $2,066.64, yet your interest will still be on the full amount of the loan."

"Do you mean that my monthly payments are going to be lower than I figured?" Joe asked happily.

"Not likely," said Stanley. "Banks figure interest the same way you do when they are lending money. The only time the true interest rate is used is when they're the ones who are paying it. In fact, your monthly payments will probably be higher than you calculated."

"How's that?" Joe asked, not so happily.

"Because most car loans are discounted," Stanley said. "On your $2,500 loan, with a finance charge of $600, the bank would pay itself the $600 interest in advance and give you only $1,900. In order to get the $2,500 you need, you'll have to borrow more."

"I don't understand," said Joe. "Or rather, I do understand that I'm going to be paying more than I thought and at a higher interest rate. But how do I calculate all this?"

"By using the same formulas the bank does," Stanley replied. "Your first question is how much more you have to borrow to get an actual $2,500 on a discounted loan. The formula is the amount you want to borrow times 100, divided by a figure which is calculated by multiplying the per cent of interest times the number of years, and subtracting it from 100. Got that?"

"No."

"Well here it is on paper:

$$\frac{\$2,500 \times 100}{100 - (8 \times 3)} = \frac{\$250,000}{76} = \$3,289.48$$

"So you have to borrow $3,289.48 in order to get the $2,500 you need. Now go ahead and figure 8 per cent of $3,289.48 multiplied by three for your three-year loan."

"I make it $789.48," Joe said.

"Right," said Stanley. "That $789.48 is immediately deducted from your $3,289.48 to give you the $2,500 you need. Now to figure out your monthly payments, just divide $3,289.48 by 36."

"My monthly payments will be $91.37," said Joe.

"Okay. Now to find the true interest rate you are paying, multiply the finance charges by 24 (twice the number of your monthly payments) and divide the result by the amount you would have borrowed without discounting ($2,500) times 37. That last figure is just your number of monthly payments plus one. Here's how it looks on paper:

$$\frac{\$789.48 \times 24}{\$2500 \times 37} = \frac{\$18,947.52}{\$92,500} = 20.48\%$$

"So you're actually paying a little over 20 per cent," Stanley concluded. "Now let's use the same formula for finding out what you would be paying if the interest was really 8 per cent. Since we don't know the finance charges but we do know the interest rate, we express the formula like so:

$$\text{Finance charges} = \frac{\$2,500 \times 37 \times .08}{24} = \$308.33$$

"The finance charges would be $308.33. Add that figure to $2,500 and divide by 36 and you will see that, at a true 8 per cent, your monthly payments would be only $78.00. Now is there anything else you'd like to know?"

"Yes. How can I get cheaper financing?"

"Ah," said Stanley. "That is the question. One way of course would be to finance a lesser amount or to reduce the period over which you repay the loan. For instance, let's take exactly the same loan at the same interest rate and see how it looks over a two-year period. First we figure out how much you would have to borrow, using the formula for discounted loans and substituting two years for three:

$$\frac{\$2,500 \times 100}{100-(8 \times 2)} = \frac{\$250,000}{84} = \$2,976.19$$

"Eight per cent of $2,976.19 times 2 (for a two-year loan) works out to $476.19, the amount of the finance charges. To find the true rate of interest, we apply the formula of interest charges times 24, divided by the real amount of the loan times number of payments plus one:

$$\frac{\$476.19 \times 24}{\$2,500 \times 25} = \frac{\$11,428.56}{\$62,500} = 18.28\%$$

"Both loans are at a nominal 8 per cent and both are discounted. The true rate for the three-year loan is 20.48 per cent, while the two-year rate is 18.28 per cent. You are paying $313.29 less for the two-year loan, however your monthly payments are naturally higher since there are less of them. You would be paying $2,976.19 divided by 24 or $124.00."

"That's more than I can afford," Joe said. "Couldn't I try for a cheaper loan somewhere else?"

"Certainly you can," Stanley replied. "You are arranging your loan through the auto dealer, which means that the bank making the loan has to give him a commission. Go directly to the bank, any bank, and you will probably be charged less for financing. Since you are a good credit risk, you may get a loan which is not discounted or be quoted a lower true interest rate.

"The financial risk has a great bearing on the rate of interest you will have to pay," Stanley went on. "If you have securities—stocks, bonds, or an insurance policy with a cash value—you can deposit them with a bank to guarantee your loan and get a substantially lower rate of interest. All the while you can still keep your life insurance policy in force or collect interest and capital gains on your securities. By not cashing your securities in, you may also be saving on income taxes if short-term gains are converted to long-term ones."

"I don't have any securities," Joe said. "As a matter of fact I'm kind of broke."

"All the more reason why you should get the best deal you can on financing charges," Stanley firmly stated. "Check a few banks and compare their rates, using the formulas I gave you. And don't forget to see if you can qualify for a loan from a credit union operated by an employee group, union, or other association to which you belong. Their rates are often the most favorable."

"Well, thanks for the advice," Joe said.

"Just one more thing. Since you don't have security for your loan, you will probably be required to buy collision insurance and also a term life insurance policy upon which the lending institution can collect if anything happens to you and the payments lapse. Consider the cost of any insurance you wouldn't otherwise buy as a part of the total cost of the loan. And don't be afraid to suggest that you can arrange insurance through your own agent if the rates you are offered seem inflated."

"I'll do that," Joe said.

"Good luck," said Stanley.

WHAT ABOUT INSURANCE?

Think of an insurance policy as a package deal where you can select a portion of one type of coverage, a little of another, some more of still another kind, and put them all together in one policy.

Liability insurance covers you against injuries to another person or damages to his property in an accident. The limits of liability insurance are often stated in three numbers, such as 20-40-5. This means that, in the event of an accident for which you are liable, the insurance company will pay *a maximum* of $20,000 for injuries to a single person, $40,000 for injuries to two or more persons, and $5,000 for property damages. Liability insurance is mandatory in some states.

Collision insurance compensates you for damages to your own car in an accident where you cannot collect from another party. Collision insurance is generally sold with a *deductible* provision. On a $100-deductible collision policy you would pay for the first $100 worth of repairs to your car in the event of an accident and the company would pay the rest. This saves the insurance company the trouble of dealing with frequent petty claims for dented fenders and such, so deductible policies are cheaper.

In addition to liability and collision, a policy may cover:

—Damages to your car from fire, theft, and natural disaster (known as *comprehensive* coverage);

—Medical payments for yourself and your passengers in accidents where you cannot collect from another party (this coverage being in addition to what your health insurance policy will pay);

—Damages you sustain in an accident where the other party (the guy who hit you) does not have any liability insurance and you are unable to collect (known as *uninsured motorists* coverage);

—Injuries or damages you cause while driving someone else's car (in which case your coverage will supplement his);

—Towing and labor involved in retrieving your car after a wreck and taking it to a garage where it can be repaired.

You are the only one who can determine which parts of the insurance package you need and how much of each. As a general guide, consider that the 10-20-5 liability limit which is the minimum in some states will not go far in covering you against an accident where substantial injuries leading to disability are involved. If you can afford it, a 50-100-10 or at least a 20-40-5 limit offers a much better safety margin.

Collision insurance, which is quite valuable when you have a new car, diminishes in value each year your car depreciates. Consider cancelling it when your car is four or five years old and worth only in the range of $400 to $800.

The cost of some of the other provisions is quite nominal. Your insurance agent should be willing to tell you what each individual provision costs (some may be automatically tossed into the basic package), and how much it will pay in the event that the bad thing you are insuring against comes to pass.

As you may know, some states are adopting "no-fault" insurance which changes the picture outlined above. States with a no-fault law (it can be mandatory or optional) set a minimum below which there is no question of liability. If the minimum is $2,000 and you are in an accident with a car driven by Joe Shickelgruber, then Joe's insurance company pays him and yours pays you—no matter who was at fault. The amounts each company will pay are determined by a legal formula. You and Joe can sue each other only when the damages are above the no-fault minimum in your state.

How can you find out what laws apply in your state? Ask your friendly insurance agent. You can ask him almost anything except where to buy insurance at cheaper rates than he can offer. That's one question you'll have to answer for yourself and it requires a little digging.

Car insurance policies are sold through auto clubs, fraternal orders, unions, professional associations, and all sorts of other groups. Their prices are often a bit lower than the best your friendly agent can offer. Of course, you no longer will have a local agent to help you fill out paperwork (reports and claims, etc.), dispense advice, and give quick service when you need a proof of insurance form to take down to the motor vehicle bureau. So balance the difference in cost against the loss of service and take your choice.

Obtaining a reasonable policy at moderate rates is really not much trouble if you're an average guy driving an average car. It's when you're an oddball who owns a 1927 Vasisdas along with a hot new Super-Kamikaze motorcycle that the trouble begins. You see, insurance companies are really a lot like bookmakers; they figure the odds and write a policy which will be profitable according to their statistics. Since their data on a 1927 Vasisdas or a Super-Kamikaze are non-existent, they will either refuse to insure you at all or else demand such a high premium that they are sure to come out ahead.

When confronted with this sort of situation, you have a couple of viable alternatives, as they say. One is to threaten your insurance agent that you will cancel your life insurance policy, the fire and theft policy on your house, and any other policy you carry with him unless he bends the arm of the insurance company he represents to reduce the premium to something approaching a reasonable rate. It's been known to work.

The other possibility is to seek out a small insurance company which specializes in covering the type of vehicle you drive, be it an antique, sports car, dune buggy, or cycle. Such companies do exist and they can offer substantially lower rates because your type of vehicle is their specialty rather than being an incomprehensible exception for which they have no statistics. Some of these specialized insurance companies advertise in enthusiast publications; others can only be found by assiduous letter-writing or word-of-mouth recommendation.

And now for the question of what to do when your insurance company takes more than a reasonable length of time to pay a claim, cancels your policy arbitrarily, or otherwise gives you reason for a legitimate gripe. Writing your congressman or Ralph Nader may give you the satisfaction of alerting people in high places, but is unlikely to bring immediate help. Better to write the agency which regulates car insurance in your state (usually the State Insurance Board or Commission) and send a copy to the insurance company which has done you dirty. Since insurance companies are closely regulated by each State Board, this is your best channel for getting quick action.

CHECKING OUT A USED CAR

Determining the condition of a used car is more a matter of sharp-eyed detective work than any special mechanical ability. With a few tools and a systematic approach, you can probably do a better job than the average mechanic whom you might hire for a few dollars to give the car a quick once-over.

Obvious clunkers can be spotted and rejected by a quick check of the externals—body, interior, major accessories, and visible parts of the engine. The next step is a thorough road test to bring out faults in the chassis, engine, or running gear. Finally, you will need a flashlight and a few test instruments to probe for hidden defects.

Since every used car has *some* flaws, don't expect the sun, moon, and the stars. Just note the items requiring repair, try to estimate the costs, and keep a running total. Take along a notebook, if necessary, and do your figuring on paper.

You should have a realistic estimate of both the cost and trouble of all necessary repairs before you make a move toward your wallet.

The Fast Check

Walk around the car at a little distance for your first general view. Sometimes this is the best way to spot a slightly crooked bumper, misaligned body part, or other indication that the car has required major sheetmetal repairs.

In these initial checks you are trying to spot a car that has been in a serious accident or has seen abnormal mileage for its age. Car dealers frequently purchase such vehicles at auctions or salvage yards. The odds are good that a rebuilt wreck will have frame or chassis damage that can never be 100 per cent repaired. A high-mileage fleet car or former taxi may look all right superficially, but wear on hundreds of little moving parts has surely taken its toll. Pass up both types of cars unless you are an enthusiastic mechanic seeking a basket case to lovingly restore.

You can also immediately spot a car that sags at the rear, or at one corner, due to weak springs. This is not too expensive to repair, but it is a fair amount of trouble and may call for welding.

Now walk right up to the car and take a close-up view. You might as well check the speedometer mileage first, since the general condition of the car should mesh with the mileage it is supposed to have gone.

Can you believe what the speedometer says? Positively maybe. Setting back a speedometer (or rather an odometer, which is the mileage recorder part) is illegal under a recent federal law. Also manufacturers have been installing some relatively tamper-proof odometers on late-model cars. The trouble is that the speedo may have been inoperative or deliberately disconnected at some point in the car's life. Either the whole speedometer or the odometer part may also have been replaced.

Misaligned numerals can indicate that an odometer has been turned back by some crude method, such as chucking the speedometer cable in an electric drill and running it backwards. The slicker jobs are impossible to detect.

The best guide to the real mileage a car has gone is the signs of wear you can spot. A methodical inspection may also reveal further indications that the car has been repaired following a serious accident.

By examining the car body carefully for areas where the paint color does not quite match, looking at the trim and glass for evidence of paint overspray, and checking such places as the interior of the hood and trunk (be sure to look under the mat) and the door jambs, you can almost always tell if a car has been repainted. Be suspicious of any car which has been re-painted after less than three years in damp or coastal areas, five years in the dry and sunny parts of the country.

When you open and close the doors, hood, and the trunk lid, be sure to note any binding, gaps, or sag. Also check for evidence of rusting around the rocker panels (beneath the doors), fender wells, rear quarter panels, and door edges. If the car has been repainted, the job may have been done to conceal evidence of body cancer.

With a small penknife blade or other pointed object you can probe sus-pect areas for signs of body cancer beneath the paint. Obviously you will have to do this discreetly and in an inconspicuous place to keep good rela-tions with the dealer.

Check the roof of the car also. If the car has seen use as a taxi or police vehicle, there may once have been a crest or rotating light on the roof. You should be able to spot evidence that holes in the roof have been filled in with body putty and covered with touch-up paint.

Inside the car look for worn or frayed upholstery, rubber ground off the brake or gas pedals, or floor mats with worn ribs. Brand-new mats or ped-als, or cheap seat covers over ragged upholstery, are also clues that the car has been flogged hard at some point in the past.

The inside roof, or headliner, is another area to check for crest holes. Also inspect the dash and under-dash area for bracket holes, indicating that a two-way radio has been removed, or bunches of disconnected wiring be-longing to some previous installation.

Look for service stickers inside the door or on the air filter can. It is sur-prising how often someone turns back an odometer without thinking to remove the stickers which show a higher mileage.

Have a friend walk around the car as you test all the lights. Blow the horn, try the radio, heater, air-conditioning (if any), and other accessories which can be checked with the car stationary.

Under the hood, check for oil leaks around the valve covers, the oil filter, and the pressure sending unit. If the engine has been steam cleaned, every-thing will be exceptionally sanitary and you won't be able to spot leaks until after you have run the car.

Pull out the dipstick and look at the oil. Oil is normally black even if it has been changed recently. However, thick, gritty or gummy oil is unhealthy. At the least, it indicates neglect. It may also be a clue that very heavy oil or syrupy additives are being used to cut oil consumption on a worn engine.

A gasoline smell around the dipstick signifies that raw gas is probably leaking out of the cylinders due to worn rings. Watery oil could be due to an internal crack in the block. If the oil seems thinner than it should be, let a drop or two fall on a hot block and see if it sizzles because of the water content in it.

Look the radiator over carefully. A new radiator, new brackets holding the radiator in place, or fresh welds, could be a sign that the car was rebuilt following a front-end collision.

Take off the radiator cap and dip your fingers in the coolant. A reddish color with particles of rust suspended in the fluid is a bad sign. Since radiators are made of rust-proof alloy, the rust indicates corrosion in the engine water-jacketing or in the water pump. This does not necessarily mean that the car is in the last stages of decay, but it is in the first stage at least.

While you are poking around under the hood, eyeball the battery, radiator hoses, fan belt, and other paraphernalia for signs of neglect. The cost of replacing some of these items is minor; however they are good indicators of general condition. The battery may have a little metal tag which is punched to show the year and month it was purchased.

At the back of the car, poke a finger in the end of the exhaust pipe. Any gummy black stuff you come up with is oil—a sign that the engine is burning too much of it and may be due or overdue for a ring job.

Squat down and look at all the tires. If the tires appear brand new, they are probably inexpensive bias-plies or they may be retreads. Retreads are now required to have the letter "R" after "DOT" (for Department of Transportation) molded in the sidewall.

Worn tires may show cupping, feathered edges, or irregular wear patterns. Since there are a number of causes for uneven wear (see Chapter 5), no quick conclusions are justified. All you can do is be alerted to the possibility of defective shocks, improper wheel balance, or tires which are out of alignment.

Tires are a major car expense, so estimate the depth of the treads (you can try the Lincoln penny test) and don't forget to check the spare.

Before road-testing the car you will also want to run a quick check on the brakes and steering. Try the steering wheel for looseness (free play), or a binding, notchy feel. Then test the brakes by holding the pedal down hard for half a minute and seeing whether or not it slowly sinks to the floor, indicating that there's a leak somewhere in the system.

If the steering is excessively loose (more than two inches of free play) or the brakes are leaking, do your insurance company a favor and ask the dealer to make repairs or adjustments before you even take the car for a spin around the block.

Loose steering may only need a simple adjustment, while leaky brakes could only require a little tightening at the pressure connections. Should the dealer be unwilling or unable to oblige, however, the problem could be more serious and you just might be stuck for a new steering box or master cylinder if you purchase the car.

A car which has passed your scrutiny up to this point is ready for road-testing. Never never buy a car which you are not able to road-test because a dealer claims he has no plates or insurance, or some such excuse. Even if you are buying from a private owner, the privilege of road-testing a car should always be granted.

On the Road

A road-test is not a joy ride or *carte blanche* to do what you like with the car. It shouldn't be a quick spin around the block either. A half-hour drive under varied conditions and with a few stops for specific tests should tell the tale. If possible, have someone follow you in your own car so that you have a ride back in the event that the test car poops out.

Ready? Put on your Uncle Tom McCahill hat and switch on the engine. Don't be surprised if it doesn't start up right away since the car may have been sitting out on the lot for a few weeks. Once started, however, it should idle smoothly, or else you will have a hard time spotting engine problems. Should the engine be really rough, you might ask the dealer to give it a fast tune-up.

As the engine warms, rev it up a bit. Do you hear a roaring noise? If so, make a note to examine the muffler and exhaust system carefully for holes.

Drive the car at slow speeds as you check to see that everything is functioning normally. Do all the dash gauges work? The ammeter should stay slightly above mid-line, indicating that the charging system is perking. Water temperature should be well below the danger zone and stay there all through your test. The oil pressure needle ought to be somewhere around mid-range also. If there are warning lights, they should not come on while you are driving or flicker erratically.

Alert the person driving behind you to be on the lookout for blue smoke coming out your exhaust system, meaning that the engine is due for an overhaul.

On a car with a manual transmission, you will want to test the gearbox and clutch. As you go up through the gears, shift precisely and note any undue play in the lever, which will require adjustment. Try shifting moderately fast. A loud "graunch" as you go into any gear indicates that the synchronizers for that gear are worn. Clicking noises as you ride along in any gear are the characteristic sound of chipped gear teeth. Another sign of worn internals is a lever that jumps out of gear as you go over rough bumps. Reject a car if the transmission is really loud or balky.

The clutch should not chatter, grab, or noticeably slip. A good test for clutch slip is to try to start the car and pull away with the handbrake on and the transmission in high gear. It should stall immediately. If it doesn't, you will probably hear and feel the clutch slip and you may get a whiff of Chanel # burnt clutch lining. Figure at least $80 for new clutch internals unless you do the job yourself, in which case the cost will be approximately half.

An automatic transmission is usually the most expensive part of a car to replace, so it should be checked carefully. Try the transmission through a series of stops and starts in all ranges. Rough or jerky shifting, a lag on the downshift when you floor the pedal, or racing of the engine before the gears change are all danger signals. Simple band adjustment or the replacement of a modulator valve might clear up the problem—then again it might not.

Listen for transmission whine or other expensive noises. For a severe test, look to see that no police cars are in the vicinity and try burning rubber from a standing start. Do this even if you are a little old lady and wouldn't think of driving that way normally. If the transmission slips or protests audibly, head back to the dealer and tell him that you'll pass this one up.

After you are done playing with the auto trans, put the lever in Park and check the transmission fluid with the engine idling. The fluid level should be up to the notch on the dipstick and the fluid itself should not smell burnt or look yellowish. A sick transmission means a sick bank account.

The brakes are next on the agenda. Try applying them at moderately high speeds and look for pull to one side, an ear-piercing screech which sounds as though the linings are worn and the rivets are grooving against a brake drum, or a spongy pedal. All are conditions requiring further investigation.

In an area where there are no other cars or kibitzers, try three or four quick stops from about 30 m.p.h., one after another. If the brakes fade drastically, chances are that they are not designed too well. You may have to go to the new metallic linings to get decent braking power. Figure another $60 or so as the cost of the conversion.

Test the handbrake on the steepest hill you can find. Handbrake adjustment is a minor item, but not one you can afford to ignore.

A simple test of the universal joints is to start, stop, and back up rapidly a few times. A distinct "clunk" as you do this means a loose universal which will have to be replaced sooner or later. The cost will only be about $20 if you do the work yourself, quite a bit more if you must pay for the labor.

Listen for grinding noises from the vicinity of the rear axle as you drive along slowly and come to an easy stop. Also chattering noises from the differential and pencil-sharpener sounds from the front wheel bearings. They all spell expensive trouble.

How has the engine been behaving as you went through all these tests? Stalling, hesitation, or a high-speed miss are all signs that a tune-up is overdue. Distinct engine knocks, clattering, piston slap, or clicking noises due to worn lifters could signify more serious trouble.

As you drive along, notice signs of heavy steering (a defective power steering pump or slipping belt, most likely), excessive wander, steering which binds on sharp turns, wheel shimmy (probably caused by out-of-balance wheels or improper front-end alignment), or a loud whine or buzz when the wheel is locked all the way over to the left or right. The latter problem is caused by a sticking valve in the power steering mechanism and can often be corrected by using an additive in the steering fluid. Other symptoms may be due to worn front-end parts, bad shocks, or ball joints with too much play.

Find a wide road without any other cars in sight and drive along with your hands momentarily off the steering wheel. This is the best way to check for excessive drift or wander in the steering.

Try going over some rough bumps or ruts in the road and see if you can feel abnormal spring rebound (jiggle). This is a sign of worn or defective shocks. If the shocks are really shot, the steering wheel may hop around in your hands as the car rebounds over fairly minor bumps or ruts.

Before you bring the car back, take time to check for oil leaks. Drive to an area where there is clean cement or fresh dirt on the ground. Park the car and leave the engine idling for a few minutes. While waiting, try the windshield wipers, washers, cigarette lighter, and other accessories you missed checking out in the lot. This is also a good time to test the efficiency of the heater or air-conditioning. In hot weather you won't get a true picture of how good the heater is, while cold weather will make it difficult to test the air-conditioner. Do the best you can and don't forget to feel for heated air at the defroster outlets and heat at the rear window if the car is fitted with an element back there.

Now move the car and examine the spot where it was standing like a soothsayer reading entrails. Oil leaks, gas leaks, or transmission fluid dribbles can be distinguished by smell, color, or rubbing the fluid in question between your fingers. A small oil spot can be expected from an older car. Any major leaks will require further inspection on your part.

At the end of your road-test, hand the car back to the dealer and go home to think things over.

Phase III

Before showing up at the dealer's again, you may want to drop in at the local library for a little research. Look through old car magazines or *Consumers Reports* for a test of the car you are considering purchasing. Read up on its weak points, service record rating, and performance. You may also want to check the car's specifications in a repair manual. This should give you an idea of what to look for when you return to the dealer's lot, clutching your tool kit and with a mean look in your eye.

By now the dealer knows what a tough customer you are, so he should just shrug his shoulders philosophically when you ask permission to make a few more tests. He might even volunteer to put the car up on a lift for you.

If no lift is available, use jack safety stands to elevate the front end. First take a general look underneath the car and see if you can spot fresh welds or any other sign that the car has had frame or chassis repairs. New undercoating may mean that the evidence was deliberately concealed.

Have someone start the engine while you check below for oil leaks. Ask the same someone to pump the brakes as you look for leaks in the lines or dribbling from inside a wheel which indicates that a wheel cylinder is defective. Have the steering wheel turned from lock to lock as you check for power steering fluid leaks.

Look at the shock absorbers to see if they are banged up or coated with moist dirt. Dented or leaky shocks must be replaced.

Spin each front wheel to check for wheel bearing bind. A wheel that mutters and growls instead of spinning freely will need new bearings.

Take a firm grasp on the steering linkage connections and see if there is noticeable play, vertically or horizontally. This is just a gross check, since you are not doing any precise measuring, but it should suffice. Look for play at the idler arm, Pitman Arm and tie-rod ends on the steering mechanism. Worn parts will have to be replaced. The ball joints can also be checked in this manner.

Pull a front wheel and check the brake linings and drums. A scored drum or less than one-sixteenth of an inch of lining means that a brake job is imminent. A deeply scored drum must be replaced.

Peel back the rubber at the ends of a wheel cylinder to look for fluid leaking past the piston. More than a few drops of fluid in the piston rubbers is an indication that the cylinders at all the wheels should be rebuilt.

Lower the front end of the car and elevate the rear on a stand. Look for leaks at the differential, sagging leaf springs, or a rusty exhaust system. Go over the length of the muffler and exhaust pipes, tapping each part with the handle of a screwdriver, for signs of rusting through. Check to see that the pipe brackets are solid and tight.

You can also pull one rear wheel to get an idea of the condition of the rear brake linings and wheel cylinder.

When you are done nosing around beneath the car, start in under the hood. Note the condition of the exhaust and intake manifolds, checking for cracks. You may want to test the entire exhaust system for leaks, as described in Chapter 3.

Check the battery with a hydrometer. Take off the distributor cap and see if the points show signs of arcing or burning, then pull all the spark plugs. Reading the plugs for signs of engine trouble is a procedure covered in Chapter 6.

Perform an engine compression test as detailed in Chapter 5. If you do not know the specifications for the particular engine you are looking at, figure that the reading on the average six- or eight-cylinder American car should run about 130 p.s.l. A deviation of more than twenty pounds per square inch in readings between cylinders is a good indication that the engine is due for an overhaul.

If your check of the spark plugs and general look at engine condition has given you the idea that something's wrong, without being able to pinpoint the problem, some readings with a vacuum gauge, as described in Chapter 5, should give you a better picture.

These final third-phase checks can be carried out by a hired mechanic if you feel diffident. He has no magical way of peering into the car's internals for a definitive answer, so he can only go through the same procedures you would, though perhaps with a more practiced eye and hand.

Before closing the deal, be sure and look at the car's serial number to see if it matches the registration. Otherwise you may be buying a hot car. The number plate on most late-model vehicles is on top of the dash, just inside the windshield.

You may be tempted to skip a few steps in checking over a used car—or all of them—because the dealer has blithely assured you that there is a "full guarantee" for thirty or sixty or even ninety days. Don't do it. Any guarantee is worth no more and no less than the skill and honesty of the dealer issuing it. Sure you can sue, but a lawsuit is one thing you don't need.

To some, the full procedure for checking out a used car may seem excessive, even paranoiac. All I can say is that a high degree of paranoia is normal, healthy behavior when it comes to buying a used car.

SELLING YOUR CAR

Should you repair your old car before selling it or offer it for sale "as is" at a reduced price? Sometimes a little fixing can raise the asking price a substantial amount. Patching holes in the exhaust system, tuning the car to run smoothly, and charging up a weak battery are examples of minor repairs that can make a big difference to a prospective buyer.

Major repairs are rarely worthwhile. Unless you have a fairly new car which has been in an accident and is worth fixing, figure on spending no more than 10 per cent of the asking price on pre-sale repairs.

Cosmetics are an important factor in used-car sales. The sprucing up should begin with a thorough cleaning which includes degreasing the engine, vacuuming the interior and trunk compartment, and polishing the chrome.

This is the time to sand down rust spots, fill small body irregularities with "Bondo," and skillfully apply touch-up paint.

Rubbing compound may be used to bring up the gloss on paint which has grown dull. Top off your work with a thorough wax job and remember to clean the wheels and whitewall tires.

Advertise the car for sale in local papers, pennysavers, or want-ad digests. Don't overlook the possibility of free advertising on bulletin boards in supermarkets, local colleges, community buildings, and business concerns.

If your car is unique enough to appeal to enthusiasts, you may want to advertise in a national publication. A "creampuff" T-series MG can fetch a premium price if advertised in a sports car magazine, while a Deuce ('32 Ford-bodied) Roadster is best marketed through a hot-rod publication.

Your car may be of special interest to enthusiasts without fitting into any clearly defined category. For instance, an old Triumph Mayflower sedan from the early post-war era is hardly a sports car or classic, yet it may be a gem to the right buyer. An eclectic publication for car nuts, such as *Hemmings Motor News* (Box 380, Bennington, Vt. 05201), which is made up solely of advertisements, might be a good place to list such a car for sale.

Your ad should be brief and to the point. State the year, make, model, engine type, and major accessories fitted. Clear abbreviations or key words such as "auto trans," "air," or "headers" will save you money and still get your message across.

You may want to include the mileage and a brief description of the car's condition. Phrases like "mint," "good ext. and fair mech.," and "parts car" are commonly employed.

Your ad is likely to get a better response if you state the asking price. Take a tip from used-car dealers and leave yourself a $25 or $50 margin to haggle with. If you are in doubt about the price to set, check a publication such as *Edmund's Used Car Prices*. This will give you the basic price which can be adjusted up or down according to the car's condition and the market in your area. In case decision-making is not your forte, you can always say something like "Best offer over $1500."

By devoting a weekend to cleaning up your car, then advertising in appropriate places, you can increase your chances of getting top-dollar. It is possible to save $300 or more on the true trade-in value a dealer would offer.

Now with your old car gone and a shiny new model sitting in your driveway, you can put away your tool kit—for a little while at least. Go on to the appendices for information on where to write away for a service manual and how to set up your own maintenance schedule.

Appendices

Appendix **A**

A Suggested Maintenance Schedule

As part of the warranty requirements on your car you must follow the maintenance schedule mandated by the manufacturer. After that it's up to you.

Nobody can establish a maintenance schedule for your car as well as you can. Even the manufacturer doesn't know your driving habits, the number of miles you put on a year, and the roads you drive over.

The following suggestions for routine maintenance are only a basis to start from. You should add the recommendations in your owner's manual. Then make a note to check any part which seems to need more frequent attention.

This schedule is based on driving 12,000 miles per year, a reasonable median. If you drive twice as many miles per year, do maintenance twice as often. If you drive 6,000 miles a year or less, base your maintenance on a three-times-a-year schedule and do major maintenance every four years instead of three. A car driven over rutted, dusty roads or with a great deal of stop-and-go driving will require maintenance more often.

During Break-in Period—After 300–500 Miles
      Change oil and filter
      Change auto transmission fluid and clean screen (or replace filter)
      Have wheels balanced and aligned
      Be sure dealer corrects any malfunctions on new car
Every Time You Stop for Gas (or at least every other time)
      Check oil and battery water
      Inspect tires and visually check inflation
      Glance under car for leaks or dragging exhaust pipe
Once a Month
      Check inflation of tires with gauge

Check fluid level in radiator
Tighten all wheel lugs
Check operation of all lights
Look under hood for leaks, frayed drive belts, or loose wires
Test for brake leaks by holding pedal down
Check steering play

Every Three Months (Summer)
Change oil and filter
Check air filter and clean if necessary
Check fluid level in master cylinder, automatic transmission, and
    power steering
Pull a wheel to check brake linings and wheel bearing grease
Clean and degrease engine
Fill windshield washer reservoir

Every Three Months (Fall)
Change oil
Check fluid level in master cylinder, automatic transmission, and
    power steering
Fill windshield washer reservoir
Check battery cells for charge and clean battery and terminal
    connectors
Lubricate body parts
Lubricate heat riser valve and automatic choke
Check anti-freeze and add if necessary
Check grease in differential, manual gearbox, and steering
Clean carburetor and linkage
Inspect underside of car for leaks and inspect exhaust system
Clean PCV valve and associated hoses
Clean oil filler cap
Grease chassis (if recommended as regular maintenance procedure
    in owner's manual)
Check shock absorbers
Perform other lubrication as recommended in owner's manual
Clean and regap spark plugs
Check point gap and file points if necessary
Check distributor parts and timing

Every Three Months (Winter)

    Change oil and filter

    Replace air filter

    Check fluid level in master cylinder, automatic transmission, and
        power steering

    Check clutch pedal play

    Check ball joint play

    Adjust headlights

    Fill windshield washer reservoir

Every Three Months (Spring)

    Change oil

    Check fluid level in master cylinder, automatic transmission, and
        power steering

    Touch up paint and clean chrome

    Check windshield wiper blades and replace if necessary

    Replace points and spark plugs

    Replace distributor cap, rotor, and condenser (if necessary); lubri-
        cate distributor

    Check point gap and timing

    Lubricate heat riser valve and automatic choke

    Lubricate speedometer cable

    Check PCV valve and hoses

    Clean oil filler cap

    Perform other lubrication as recommended in owner's manual

    Check hoses, drive belts, and wiring for visible wear

    Check shock absorbers

    Fill windshield washer reservoir

Every Second Year

    Repack front wheel bearings with grease

    Replace in-line fuel filter

    Replace automatic transmission fluid and clean screen (or replace
        filter)

    Replace vacuum modulator valve if shifting is rough

    Replace filter on air injection system

    Add can of rust inhibitor and water pump lubricant to cooling sys-
        tem

Have wheel balance and alignment and front end wear checked after purchasing new tires

Every Third Year

Replace radiator and heater hoses

Drain cooling system and flush; put in new coolant

Replace radiator thermostat

Replace worn or frayed drive belts

Replace spark plug wires and coil wires

In addition, be alert for any developing problems. Keep your eyes and ears open. Replace worn brake linings, bad shock absorbers, weak battery, slipping clutch plate, worn ball joints, or other suspension and steering parts promptly.

Appendix **B**

The Tool Kit

The items in the $75 tool kit listed here are the essentials for small repair and maintenance jobs and will also serve you well as a trunk compartment emergency kit when combined with a selection of items from the Spare Parts and Miscellaneous list.

The $125 supplementary tool kit will see you through many advanced repairs, with the exception of those requiring some of the special tools mentioned in Chapter 10, and will make some maintenance operations simpler. Some of the items in this kit would be appropriate to carry on long trips or expeditions in remoter areas. Carry as much as you can afford and stop only when the rear springs of your car start noticeably sagging.

Foreign-car owners will want to substitute metric wrenches for American (SAE) sizes. Some of the older British cars used Whitworth bolts and require combination and socket wrenches of a still different size.

American cars manufactured abroad, such as the Ford Courier and Chevrolet LUV mini-pickups imported from Japan, use metric sizes. A few cars, such as Volvos and some English cars, have imported components and may require both SAE and metric (or Metric and Whitworth) sized wrenches. Check the service manual for your car if there is any doubt about what wrenches to buy.

There is a degree of correspondence between all three systems. Some wrenches made for one type of bolt size will fit another—or almost fit. The latter is more often the case and, since near fits cause rounded-off bolts and nuts, you will have to invest in extra sets of wrenches if you regularly work on two dissimilar cars. For occasional tinkering you can get by with substitutions and adjustable wrenches.

Don't forget that cars which require metric or Whitworth sized wrenches also will necessitate your stocking appropriate bolt and nut sizes in your spare parts kit. In addition, screw threads are sized differently on many foreign cars. Although you should be aware of the problem, don't worry about it unduly. When you need special bolts for major repairs or rebuilding, such items generally are a part of rebuild kits. Also you will obtain odd-sized bolts and screws by stripping old assemblies of useable spare parts.

Along with tools and parts, I have listed items which are necessary during the winter in areas where it snows. Another small list comprises two essential safety items which should be supplied with all new cars, but are not available with any so far as I know. These are:

"Cat's eye" reflector kit	$7.00
Fire extinguisher	about $15.00

The reflector kit consists of three reflectors on stands which should be placed behind your car whenever you are forced to make an emergency stop on the highway or close to the shoulder (see Chapter 8). Flares will also do but they do not last more than fifteen or twenty minutes apiece. A reflector kit of this nature is sold by Western Auto Stores and many other suppliers. You can also make a pretty fair one yourself out of luminescent tape, small squares of masonite or heavy cardboard, and coat hanger wire bent to form a stand.

The fire extinguisher should be a Type BC (B is necessary for gasoline fires, C is the rating for effectiveness against electrical fires) dry powder type. It should have Underwriters' Laboratories (UL) approval on the label and be of at least two-and-a-half pound capacity.

The fire extinguisher should have a built-in gauge which you can check regularly. Once you have used the extinguisher, take it to a dealer and have it recharged. (You can locate such services through the Yellow Pages.) It's a good idea to have the extinguisher tested by a dealer every five years.

For ready access in emergencies, the extinguisher should be mounted on brackets within the driver's compartment. Brackets can be purchased with the extinguisher and a convenient mount place is somewhere underneath the dash where neither the driver nor passenger is likely to strike it if pitched forward in an auto accident or emergency stop.

$75 TOOL KIT

Combination wrenches, ⅜″ to ¾″ by sixteenths	$10.00
Socket wrench set (⅜″ drive with ratchet handle, spark plug socket, 3″ extension and sockets from ⅜″ to ¾″)	$15.00
Water pump pliers	$ 4.00
Vise grips	$ 3.00
Crimping pliers (with wire cutter and stripper)	$ 3.00
1½ ton scissors jack	$ 7.00
Lug wrench (cross type)	$ 3.00
Small, medium, and large screwdrivers	$ 4.00
Small and medium Phillips screwdrivers	$ 2.50
Jumper cables (12′ copper core cables, at least six gauge)	$12.00
Flashlight	$ 2.00
Shovel (Use the one you have around the house.)	
Oil can spout	$ 1.50
Oil filter wrench	$ 1.00
Feeler gauges (flat and wire type)	$ 1.50
Wheel chocks (Make them yourself or use stones.)	
Tool box (Scrounge any appropriate container. An old Army ammo box is fine.)	
Hammer (small ball peen)	$ 2.00
Tire pressure gauge	$ 1.50
Pocketknife (You already have one, don't you?)	
Drift punch and small cold chisel	$ 2.00
	$75.00

COLD WEATHER KIT

Chains (if needed, or studded tires)
Windshield de-icer (Hot Melt is best)
Windshield scraper (sturdy hard-rubber one)
Old broom (for sweeping off snow)
Spare chain links
Fog cloth

$125 SUPPLEMENTAL TOOL KIT

Hi-lift jack	$26.00
Tow rope or cable	$ 6.00
Jack safety stands (2)	$ 8.00
Hacksaw	$ 4.00
Needle-nose pliers	$ 3.00
Socket wrench extension 6″, universal swivel adapter, ⅜″ to ¼″ adapter, ¼″ drive sockets sized ⁵/₁₆″, ¼″, ¹¹/₃₂″	$ 9.00
Set of open-end ignition wrenches	$ 8.00
Battery hydrometer	$ 2.00
Anti-freeze tester	$ 2.00
6″ adjustable wrench (crescent type)	$ 2.50
12″ adjustable wrench (crescent type)	$ 6.00
Stubby screwdrivers (regular and Phillips)	$ 3.00
Offset screwdriver	$ 1.50
Set of Allen wrenches	$ 2.50
Timing light (or Compu-Dwell)	$10.00
Wire brush	$ 2.00
Low voltage continuity tester	$ 3.00
Transmission-differential wrench	$ 2.00
Compression gauge	$ 5.00
Vacuum gauge	$ 5.00
Grease gun	$ 4.00
Battery clamp lifter	$ 2.00
Valve stem wrench	$ .50
Magnetic screw starter	$ 1.50
Cheater bar	$ 1.50
Combination file	$ 2.00
Point file and FlexStones	$ 1.50
Oil can	$ 1.50
	$125.00

SPARE PARTS AND MISCELLANEOUS

Ignition points   Light machine oil
Spark plugs   Silicone spray

Condenser
Distributor cap and rotor
Fan belt
Power steering or air conditioner belt
Sealed-beam headlight
Small light bulbs (assortment)
Tire repair kit
Assorted fuses
Windshield wiper blades
Gas can and gas
Can of oil
Anti-freeze/water mixture
Brake fluid
Gasket sealant
Gasket material
Penetrating oil

Epoxy kit
Hose clamps
Solderless terminals
Tire valve cores
Mechanic's wire
14-gauge electric wire
Plastic tape
Duct tape
Fuel line tubing
Nuts and bolts
Lock nuts and lock washers
Cotter pins
Rags
Boards (for prying, jack base, etc.)
Spare radiator cap
Thermostat

Appendix **C**

Sources for Tools

Silvio Hardware Co.
107–109 Walnut St.
Philadelphia, Pa. 19106 (catalog 50¢)
   (*Excellent general selection of quality brands*)

U. S. General Supply Corp.
100 General Place
Jericho, N.Y. 11753 (catalog $1)
   (*Some cheap tools, but good ones also. General selection*)

Snap-On Tools
Kenosha, Wisc. 53140 (catalog $1)
   (*Top quality, comprehensive, and expensive. For the professional mechanic or wealthy enthusiast*)

S-K Tools
Dresser Industries
3201 N. Wolf Rd.
Franklin Park, Ill. 60131
   (*Fine quality line at moderate-to-high prices*)

Mac Tools
Sabina, Ohio 45169
   (*Excellent value quality tools*)

Crescent Tools
200 Harrison
Jamestown, N.Y. 14701
    (*Good, medium-priced line by the people the crescent wrench was named after*)

Husky and New Britain Tools
South St.
P.O. Box 1320
New Britain, Conn. 06050
    (*Medium-to-high prices in a top line*)

Proto Tools
Pendleton Tools Industries
2209 S. Santa Fe Ave.
Los Angeles, Calif. 90058
    (*High prices and quality. A real professional line*)

K-D Tools
K-D Manuf. Co.
Lancaster, Pa. 17604
    (*All types of specialty tools*)

Owatonna Tool Co.
Owatonna, Minn. 55060
    (*Also for specialized tools*)

J. C. Whitney & Co.
1917–19 Archer Ave.
Chicago, Ill. 60616
    (*The old favorite for tools and parts. Many specialty tools but regular lines of wrenches, etc. are mainly cheap sets*)

Sears-Roebuck & Co.
303 East Ohio St.
Chicago, Ill. 60611
    (*Or your local store. The Craftsman line is Sears's best and prices are good*)

Montgomery-Ward
619 West Chicago Ave.
Chicago, Ill. 60607
(*Or your local store. Powr-Kraft is the top line Wards handles*)

You can get a tire pressure gauge, a tread depth gauge, four valve caps, and a booklet on tire care for $1.50 by writing to Tire Safety Council, P.O. Box 726, New York, N.Y. 10010.

Appendix **D**

Spare Parts

If you live in or near a metropolitan area, most parts should be available locally. Nevertheless, there are a number of reasons for shopping by mail. You may need an odd part, a piece of speed equipment, or a part for an older or foreign car.

It would take an encyclopedia to list all the makers of high-performance parts, accessories, and dress-up equipment for all types of automobiles. Sears, Ward, and J. C. Whitney (addresses given in Appendix C) are the giants in the field. Sears has special catalogs for foreign cars parts and for dune buggy and off-road vehicle accessories. The Whitney catalog is very comprehensive and is a good standard for comparing prices.

Some other suppliers are:

Warshawsky & Co.
1900–24 So. State St.
Chicago, Illinois 60616
>   (*The J. C. Whitney catalog under a different name*)

Spiegel, Inc.
Chicago, Illinois 60609
>   (*Comprehensive catalog free*)

BAP/Geon
East Coast: P.O. Box 2000
>   Woodbury, N.Y. 11797
West Coast: P.O. Box 5000
>   Compton, Calif. 90221
>   (*Parts for most foreign cars. Write for list of local distributors*)

Honest Charley Speed Shop
108 Honest St.
Chattanooga, Tenn. 37411
(All kinds of goodies from Honest Charley himself. Free catalog)

Dick Cepek
P.O. Box 1181
9201 California Ave.
South Gate, Calif. 90280
(Good source for Hi-lift jacks and much else that belongs in an emergency tool kit. Free catalog)

EMPI
P.O. Box 1120
Riverside, Calif. 92502
(Everything for the VW. Catalog $1)

Moon Equipment Co.
10820 S. Norwalk Blvd.
Santa Fe Springs, Calif. 90670
(Performance parts and speed tools. Catalog $2)

Con-Ferr
300 N. Victory Blvd.
Burbank, Calif. 91502
(Everything for the four-wheel-driver. Catalog $1.50)

Wilco
P.O. Box 1128
Rochester, N.Y. 14603
(Sports car and rally accessories. Catalog 50¢)

Vilem B. Haan, Inc.
10305 Santa Monica Blvd.
Los Angeles, Calif. 90025
(Racing and rally equipment mostly. Some more utilitarian items. Free catalog)

MG Mitten
P.O. Box 4156 Catalina Station
Pasadena, Calif. 91106
> *(Tools, accessories, and miscellaneous goodies with a sporting flavor. Free catalog)*

Appendix **E**

Service Manuals

There are three types of service manuals. One is the official factory manual which usually covers all models of a single make and is revised annually. The factory manuals are quite comprehensive but not always easy to understand. They are designed for authorized repair centers with factory-supplied tools and written for experienced mechanics.

A number of publishers put out service and repair manuals for the more popular foreign and American cars. These are almost always oriented toward readers who are not professional mechanics and are therefore easier to use. However, some are infrequently revised or try to cover too much ground in a relatively slim volume.

There are also car repair annuals covering all major repairs on vehicles in a certain category (American cars, foreign cars, trucks, etc.). These are big, expensive books, but they do give a great deal of information.

The listing here begins with sources for factory manuals, then goes on to publishers who put out annuals and a line of repair books for individual makes of cars.

American Motors Corp.
Automotive Customer Relations Section
14250 Plymouth Rd.
Detroit, Mich. 48232
 (American Motors cars including Ramblers and Jeeps)

Buick Motor Division
Service Publications Dept.
General Motors Corp.
Flint, Mich. 48550
 (Includes Opel Kadett)

Cadillac Motor Car Division
Service Publications Dept.
General Motors Corp.
Detroit, Mich. 48232

Helm, Inc.
2550 East Grand Blvd.
Detroit, Mich. 48211
>   (*Official service manuals for all Chevrolets and Fisher Body Manuals*)

Chrysler-Plymouth Division
Service Dept.
Chrysler Motors Corp.
P.O. Box 1658
Detroit, Mich. 48231
>   (*Chrysler, Imperial, Plymouth, and Valiant*)

Dodge Division
Service Dept.
Chrysler Motors Corp.
P.O. Box 1259
Detroit, Mich. 48231

Ford Service Publications
P.O. Box 7750
Detroit, Mich. 48207
>   (*All Ford products*)

Oldsmobile Division
Service Dept.
General Motors Corp.
Lansing, Mich. 48921

Drake Printing Co.
1000 West Eight Mile Rd.
Ferndale, Mich. 48220
>   (*All Pontiac products*)

Nissan Motor Corp.
137 East Alondra Blvd.
Gardena, Calif. 90247
    (Datsun cars)

Fiat Roosevelt Motors
532 Sylvan Ave.
Englewood Cliffs, N.J. 07632
    (All Fiats)

British Motor Car Distributors
1200 Van Ness Ave.
San Francisco, Calif. 94109
    (Austin, Jaguar, MG, Triumph)

Mazda Motors of America, Inc.
3040 E. Ana St.
Compton, Calif. 90221

Mercedes-Benz of North America, Inc.
One Mercedes Drive
Montvale, N.J. 07645

Peugeot, Inc.
107–40 Queens Blvd.
Forest Hills, N.Y. 11375

Porsche Audi
600 Sylvan Ave.
Englewood Cliffs, N.J. 07632

Renault West Service Center
702 West 190th St.
Gardena, Calif. 90247

Saab Motors Inc.
100 Waterfront St.
New Haven, Conn. 06506

Toyota Motor Distributors, Inc.
2055 West 190th St.
P.O. Box 2991
Torrance, Calif. 90501

Volkswagen of America, Inc.
Englewood Cliffs, N.J. 07632

Volvo Western Distributing Co., Inc.
1955 190th St.
Torrance, Calif. 90501

Autopress Ltd.
Bennett Rd.
Brighton, BN2 5JG, Great Britain
   (*Most English cars and some continental makes*)

Floyd Clymer Publications
222 North Virgil Ave.
Los Angeles, Calif. 90004
   (*American cars, foreign cars, and motorcycles*)

Robert Bentley, Inc.
872 Massachusetts Ave.
Cambridge, Mass. 02139
   (*British and continental cars, technical books*)

John Muir Publications
P.O. Box 613
Santa Fe, N.M. 07501
   ("How to Keep Your Volkswagen Alive: A Manual of Step-by-Step
   Repair for the Compleat Idiot," *excellent book on the VW Beetle
   and Transporter*)

Carbooks, Inc.
2628 Atlantic Ave.
Brooklyn, N.Y. 11207
   (*Distributor for many service manuals and repair books*)

Autobooks
2900 R Magnolia Blvd.
Burbank, Calif. 91503
> (*Distributor for KGM, Intereurope, and other foreign and American car repair manuals*)

Classic Motorbooks, Inc.
1415/X1 West 35th St.
Minneapolis, Minn. 55408
> (*Distributor for racing, repair, and technical auto books*)

Motor
250 West 55th St.
N.Y.C., N.Y. 10019
> (Motor's Auto Repair Manual—*probably the best of the big annuals*—Motor's Truck Repair Manual, *many technical books*)

Chilton Book Co.
401 Walnut St.
Philadelphia, Pa. 19106
> (Chilton's Auto Repair Manual, Chilton's Auto Repair Manual 1940–1953, Chilton's Auto Repair Manual 1954–1963, Glenn's Foreign Car Repair Manual, *repair and tune-up guides for individual makes of cars*)

National Automotive Service
P.O. Box 10465
San Diego, Calif. 92110
> (*Technical books, annual repair book for American cars*)

Petersen Publishing Co.
8490 Sunset Blvd.
Los Angeles, Calif. 90069
> (*Series of "Basic" books, such as* Basic Ignition and Electrical Systems. *Good theory and practical information*)

Index

Accidents:
  fires, 290-291
  precautions against, 290-291

Air Conditioning, 66-68
  checking refrigerant level, 67
  checking the system, 67-68
  explanation of, 66
  installing add-on type, 68
  testing electrical load of, 68
  types of, 68

Air Cooled Engines, 62-63
  description, 62-63
  worn fan belt, 63

Air Distribution Manifold (Fuel Injection):
  description, 101

Air Filter, 16-18
  cleaning, 17-18
  changing, 17
  clogging, 16
  when to replace, 17

Air Injection, 100-103
  checking system, 102
  explanation of, 100
  servicing, 101

types of filters in, 101-102
valves in system, 101-103

Alternators:
  alternators vs. generators, 111
  description of, 12-14, 143-144
  diodes, 144
  installing, 147
  maintenance, 147
  removal of, 147
  see also *Generators*

Anti-Freeze:
  testing, 53
  use of, 19-20

Anti-Stall Dashpot:
  explanation of, 72-73

Auto Repair (sources for):
  see *Garages*

Auto Storage:
  precautions for, 225-226

Automatic Transmission:
  care of, 209
  checking fluid level, 27, 213
  cleaning, 210
  draining, 209
  explanation of, 22-25

materials for, 191-192
procedure for, 190-191
valves in, 190

Brake Linings:
checking for wear, 365
description, 177
inspection of, 178-179
installing, 194-197
purchasing, 194-195
removing, 194-197
types of, 177-178

Brake Master Cylinder:
bleeding, 190-193
description, 28-29
finding leaks, 286
location of, 29
loose cylinder, 286
rebuilding, 186
removing, 186

Brake Proportioning Valves:
checking, 193
explanation of, 193

Brakes:
adjusting, see *Brake Adjustment*
bleeding master cylinder, 190-193
checking, 29
checking brakes on used cars, 361
checking for leaks in wheel cylinder, 190-193
cleaning, lubricating, and rebuilding wheel cylinder, 189-190
hydraulic, 28
light switches, see *Brake Light
Switches*
proportioning valves, see *Brake
Proportioning Valves*
relining, 194-197

repairing leaks in brakeline, 286-
287
shoes for, 28
some defects and causes, 363
tools for relining, 195-196
see also *Braking System*

Braking System:
explanation of, 28-30
leaks in, 183
parts of, 28-30
replacing brakeline, 184

Brazing:
see *Welding*

Bulbs (Light):
for flashlights, 281
for foreign cars, 114
how to replace, 113-114
numbering system for, 114
tools for replacing, 113
types of, 113-114

Bumper Jacks:
see *Jacks*

Burglar Prevention:
see *Theft Prevention*

Buying Abroad:
making arrangements, 350
possible savings, 350

Camshaft:
description, 4
in valve adjustments, 252
see also *Engine*

Carburetor:
adjustments for, 77

normal pressure, 19
parts of, 18-20
radiator cap, 19
refrigerant in, 67
tools for cleaning, 53
when and how to flush, 59-61

Cost of Operating and Owning a
    Car:
    depreciation, 334-335
    expenses, 335-337
    figuring daily costs, 337
    financing, 334
    mileage, 334

Crankcase:
    description, 98
    removal of oil pan, 323

Crankcase Breather Cap:
    cleaning, 100
    description of, 100

Crankshaft:
    description of, 4
    see also *Engine*

Cylinders:
    see *Engine Cylinders*

Dashboard Gauges:
    see *Gauges, Dashboard*

Deadman:
    see *Winching*

Dealers:
    see *Automobile Dealers*

Defroster:
    explanation, 66

Differential Gears:
    explanation of, 25-26

Disassembly Procedures (General),
    316-318

Disc Brakes:
    bleeding, 197-198
    explanation of, 197
    fluid for, 30
    replacing pads on, 198

Distributor:
    breakerless, 260
    CD system, 260
    checking vacuum advance, see
        Distributor Vacuum Advance
    dual points, 243, 258
    parts of, 8, 228
    removing, 238-239
    removing rotor, 239
    replacing, 244
    replacing rotor, 239, 244-246
    rotor, 228-229
    stationary rotor (troubleshooting),
        269

Distributor Cap:
    inspecting, 247
    removal of, 8, 238
    repairing cracks in, 270
    replacing, 247

Distributor Points:
    adjustments for GM and AMC
        V-8 cars, 242
    aligning, 241
    cleaning, 241
    explanation of, 227-229
    gapping, 230, 242
    replacing, 240-241
    setting, 242

Distributor Vacuum Advance:
  checking, 246
  when timing ignition, 247

Drive-Shaft:
  correcting minor imbalance, 326-328

Electric Motors:
  explanation, 140-141

Electrical System:
  how to clean external components, 106-107
  jumping ignition switch, 287
  switches, 123-125
  testing and replacing switches, 124-125
  types of relays, 144
  voltage, 271
  see also *Circuit Breakers*

Electrical Wiring:
  see *Wiring*

Elevating the Car:
  equipment needed for, 32
  for examining frame and chassis, 365
  for shock inspection, 32
  safety precautions, 32

Emergency Brake:
  see *Hand Brake*

Engine:
  cleaning procedure, 106-107
  cost of ring job, 318

cost of valve job, 320-323
cycles, 4
disassembly procedures, 318-320
explanation of, 2-6
fails to start, 267-275
lubrication, 5-6
normal compression, 366
parts of, 2-6
prolonging life of, 213
steam cleaning, 106
Wankel, see *Wankel Engine*
water leaks in, 282
see also *Camshaft, Crankshaft, Flywheel, Pistons, Valves*

Engine Cylinders:
  bad head gasket, 267
  draining, 267
  explanation of, 2

Engine Valves, 252-256
  adjusting on VW's, 253-254
  adjustment, 252-256
  adjustment when reinstalling head, 320
  checking, 254
  description, 252-253
  specifications of, 253
  types, 4
  see also *Engine*

European Purchase of New Cars:
  see *Buying Abroad*

Exhaust Manifold Heat Valve:
  see *Heat Riser Valve*

Exhaust Pipes:
  brackets for, 36
  repairing in emergencies, 290

Exhaust System:
    checking for rust and tightness, 365
    cleaning, 105
    inspection of, 36, 103-104
    repairing, 104
    replacing, 104
    testing for leaks in, 105
    tools for repairing, 104-105

Factory Repair Manuals:
    see *Service Manuals*

Fan Belts:
    checking tightness of, 14
    description of, 12
    for Corvair, VW, Porsche, 63-64
    replacing, 14

Financing, 347-354
    cost of insurance and loans, 353-354
    dubious gimmicks, 347
    formulas for figuring interest and charges, 352-353
    guaranteed loans, 353-354
    where to get a loan, 353

Flashers, 118-119
    explanation of, 118-119
    locations of, 119
    replacing, 119

Flat Rates, 304-305
    flat rate manuals, 305
    system, 304-305

Flywheel:
    description of, 20, 206
    see also *Engine*

Foreign Cars:
    advantages and disadvantages, 340

Freeze Plugs:
    replacing, 59

Freon, 12
    in air-conditioning system, 66

Front Wheel Bearings, 179-181
    checking, 180
    checking binds, 365
    how to tighten, 181
    repacking, 179-180
    servicing, 179-181
    strange noises in, 363

Front Wheel Drive Cars, 26

Fuel Injection (Electronic), 93-97
    advantages, 95
    Bosch, 95
    description of, 93
    location of injectors, 95
    parts of system, 93-96
    servicing, 96-97

Fuel Injection (Mechanical):
    explanation of, 94-95

Fuel Pump:
    description of, 16, 86
    electrical type, 16, 86, 88
    in Chevy Vega and Jaguar, 86
    mechanical, 86, 88
    problems with, 87
    replacing, 88
    see also *Fuel System*

Fuel System:
    checking for gas in carburetor, 269

on V-8 engines, 92
parts of, 91

Heater Cables:
  repairing, 65

Heaters, 65-66
  "bleeding" the air, 65
  causes of failure, 65-66
  locating and inspecting blower in,
    66
  operation of, 65
  repairing, 65

Hemmings Motor News, 367

Hollander Manual, 70

Horns, 127-128
  adjusting, 128
  repairing, 127-128
  troubleshooting, 127-128

Hose Clamps, 55-56
  removal of, 56
  types, 55

Hoses (Heater and Radiator):
  inspecting, 53-54
  loose hose clamps, 281-282
  repairing leaks in, 281-282
  replacing, 54

Hydrometer, Anti-Freeze:
  determining percent of anti-freeze
    in radiator coolant, 53

Hydrometer, Battery:
  cost of, 107
  reading, 111
  use of, 109-111

Ignition System:
  CD ignition, 260
  checking for spark, see *Spark Plugs*
  coil, 6-7, 227
  distributor, 6
  distributor rotor, 7
  explanation of, 6-8, 227-231
  faulty resistor, see *Wiring*
  jumping ignition switch, 287
  magneto, 260
  moisture in, 273
  points, 7, 227
  timing in, 230-231, 248-250, 269
  transistorized, 260
  waterproofing, 148-149

Insurance, 354-357
  buying, 354-357
  coverage, 355-356
  deductible provisions, 355
  insuring exotic vehicles, 356-357
  liability and collision, 354
  liability limits, discussion of, 355
  medical payments, 355
  "no-fault," 355-356
  registering a complaint, 357
  state insurance boards, 357
  uninsured motorist coverage, 355
  use of statistics in figuring premi-
    ums, 356

Jacks:
  additional support on soft ground,
    294-295
  cost of, 32
  hydraulic, 33, 310
  Hi-Lift jack for freeing stuck car,
    294
  placement of, 33
  scissors, 33, 310
  techniques with Hi-Lift jack, 295-
    297

Junkyards:
see *Parts, Obtaining*

Labor Costs:
see *Flat Rates*

Leaks:
how to repair, 50, 284-286
in wheel cylinder, 365
kits for repairing, 50
testing for leaks in oil, gas, brakes,
and transmission, 364-365

Leasing, 241

Light Bulbs:
see *Bulbs, Light*

Limited Slip Differentials:
advantages in gaining traction, 293

Lubricants:
types of, 52-53
see also *Oil*

Lubrication, 46-50
of door latch, 46-47
of hood latch, 46-47
of operating parts, 48-50
of U-Joints, 326
of various body parts, 46-48
see also *Grease*

Lug Nuts, 35-36
tightening and loosening of, 35-36

Machine Shops:
engine head work, 318

Manifolds, Intake and Exhaust:
explanation of, 90

repairing, 90
see also *Heat Riser Valve*

Manual Transmission:
countershaft, 20
explanation of, 20-22
jammed, 284
parts of, 20-22
signs of worn transmission, 362
synchromesh, 22
tools for repairing, 324

Mechanical Fuel Injection:
see *Fuel Injection, Mechanical*

Monroney Sticker, 343

Motor's Automobile Troubleshooter,
315

Motor's Repair Manual, 85

New Car Prices, 344-345
dealer discounts, factors in, 344
establishing a price, 345
mark-ups, 344
see also *Monroney Sticker, Price
Guides, United Auto Brokers*

Noises, Mechanical:
strange mechanical noises, 363
troubleshooting by ear, 278

Oil, 39-42
changing, 40-41
checking level, 6
for cleaning carburetors, 81
how to drain, 40
inspecting as clue to car condition, 360

quality, 39-40
types of, 39
weights, 40
when to change, 214
where to dispose, 42

Oil Filler Cap:
see *Crankcase Breather Cap*

Oil Filters, 40-41
changing, 40-41
tools for changing, 41
types of, 40

Overheating, 281-283
adding water, 283
by a plugged radiator, 283
by faulty thermostat, 282
by loose or missing fan belt, 282
by loose radiator cap, 282
caused by leaks, 281-282
in air-cooled cars, 65

Painting:
comparison of enamel and acrylic, 331
masking, 329
patching, 329
priming, 330
procedure, 220
sanding, 329-330
spray gun techniques, 330
temperature for spraying, 330
use of rubbing compound, 367

Parts, Obtaining:
by mail, 308-309
junk yards, 308
parts houses, 69-70
rebuilding kits, 70-71
rebuilt parts, 307-308
used, 308

PCV System, 97-100
clogged PCV valve, 98
explanation of, 97
PCV valve, 98
repair and cleaning, 98-99
servicing, 97-98
testing PCV valve, 100
when to check, 98

Pistons:
description, 2
see also *Engine*

Pitman Arm:
on steering gear, 173

Points:
see *Distributor Points*

Power Steering, 200-203
"bleeding," 200
checking for leaks in, 203
maintenance of, 200-203
noises in, 363
repairing, 201
vacuum boosters for, 202

Price Guides, 345

Prices:
see *New Car Prices*

Push-Starting:
techniques for, 263-264
with auto transmissions, 264

Radiator:
cap, 19
coolant level, 19
inspecting as clue to car condition, 360

Radiator (*continued*)
pressure flush, 61
see also *Cooling System*, and *Over-heating*

Radios, 135-136
curing static, 135-136
repairing, 135
replacing antennas, 135
whining and buzzing, 136

Rear Axle:
disassembling, 328
servicing, 328-329
strange noises in, 363

Rebuilt Parts
see *Parts, Obtaining*

Refrigerant:
see *Air Conditioning*

Resale Values:
factors affecting, 338

Ring Job:
see *Engine*

Road-Testing a Used Car:
procedure for, 361-364

Rocking:
as a technique for gaining traction, 293-294

Selling A Car, 367-368
content of advertisement, 367-368
cosmetic preparation, 367
setting a price, 368
where to advertise, 367

Serial Number:
location of, 366

Service and Repair Stations:
see *Garages*

Service Manuals, 315
factory manuals, 309
manuals published in England, 309

Shocks, 31-34
cost of, 32
for heavy loads, 32
how to replace, 160-163
inspection of, 32, 34, 161
signs of bad shocks, 364
testing of, 31-32
when to replace, 31-32

Shop Manuals:
see *Service Manuals*

Solenoid:
see *Starter Solenoid*

Snow Tires:
chains for, see *Chains*
types of, 157-158
when to put on, 157
with studs, 158

Spark Plugs:
checking for spark, 269
cleaning, 236, 271
gapping, 230, 237-238
inspection of, 236
removing, 234-236
replacing, 247-248

Speedometer, 50-52
cables, 50-52
checking, 52
head, 52
resetting, 358

Wheel Bearings:
see *Front Wheel Bearings, Rear Axle*

Wheel Cylinder:
see *Brakes*

Whitney, J. C., 305

Winching, 297-299
Lug-All winch, 299
power winches, 299
techniques, 297-298
with a deadman, 298-299
with hand winch, 297

Windshield Washers:
checking, 133
replacing, 133
types of, 133

Windshield Wipers:
electrical, 131

faulty motor and wiper transmission, 131
in Lincoln Continentals, 133
operation of, 130
repairing motor for, 152
types of, 130
vacuum-operated, 131

Wiring:
broken or frayed wires, 120
equipment for repairing, 120
repairing, 120-122
resistance, 273
soldering, 123
splicing, 123
testing, 121
tracing wire to origin, 120
troubleshooting with test probe, 120
use of solderless terminals, 123
wiring diagrams, 121-122

Wreckers:
see *Parts, Obtaining*